Confluence of Multidisciplinary Sciences for Polymer Joining

S. Arungalai Vendan · M. Natesh
Akhil Garg · Liang Gao

Confluence of Multidisciplinary Sciences for Polymer Joining

Springer

S. Arungalai Vendan
Electronics and Communication Engineering
School of Engineering
Dayananda Sagar University
Bangalore, India

M. Natesh
VIT University
Vellore, Tamil Nadu, India

Akhil Garg
Intelligent Manufacturing Key
Laboratory of Ministry of Education
Shantou University
Shantou, Guangdong, China

Liang Gao
State Key Laboratory of Digital
Manufacturing Equipment
and Technology, School of Mechanical
Science and Engineering
Huazhong University of Science
and Technology
Wuhan, China

ISBN 978-981-13-0625-9 ISBN 978-981-13-0626-6 (eBook)
https://doi.org/10.1007/978-981-13-0626-6

Library of Congress Control Number: 2018957046

This Springer imprint is published by the registered company Springer Nature Singapore Pte Ltd.
The registered company address is: 152 Beach Road, #21-01/04 Gateway East, Singapore 189721, Singapore

About This Book

This book adopts a holistic approach to present a course on plastic joining for researchers. Basic terminologies of plastics and the manufacturing processes deployed for joining them are discussed to benefit learners belonging to all strata of academics, industries and research sectors. The evolution of plastic joining and the advancements over the last three decades as reported in literatures have been comprehensively presented. Interesting observations, technical glitches, unusual material behaviour, inexplicable process behaviour and the challenges encountered during the research voyage of plastic joining have been briefed. This is followed by case studies on polymer joining for various combinations of thermoplastic materials espoused for components in automobile sectors. The discussion involves raw material purchase, design procedures, precautions to be undertaken, experimentation for sample preparation, welding trials, process parametric analysis and testing and characterization of polymer weld samples. To understand, examine, analyse and infer data of polymer welding in detail demands versatile knowledge on various streams of science for which it is imperative to record the signals without distortions at various stages of transfer. With the advent of sophisticated sensors/transducers and data acquisition system, several new coordinates of science for result interpretations have originated which are embedded in the case studies. The contents of this book are expected to systematically disseminate the knowledge of science underlying polymer joining to engineers.

Contents

About the Authors

Dr. S. Arungalai Vendan is presently Associate Professor in the School of Engineering at Dayananda Sagar University, Bangalore. Previously, he was a faculty member in Industrial Automation and Instrumentation Division at VIT Vellore. He undertook research on advanced welding processes since 2006. He received his Ph.D. from the National Institute of Technology (Institute of National Importance), Tiruchirappalli, India, in 2010. He has received several fellowships and awards for his technical contributions by various government and private organizations. He has successfully completed numerous government-funded research projects and industrial consultancy tasks and has published more than 80 research papers in reputed international journals and conference proceedings. He has associations with top manufacturing industries and research and development centers under various capacities. His research interest mainly focuses on the interdisciplinary science underlying welding which includes the confluence of terminologies from electrical/mechanical/metallurgical materials and magnetic streams.

Mr. M. Natesh is a Research Associate at the School of Mechanical Engineering, VIT University, Vellore, India. He has been exploring the research potentials of advanced manufacturing methods for engineering materials since 2009. He has been involved in developing mathematical/numerical modeling for parametric assessments and has contributed to the development of prototype experimental models. Currently, he is pursuing research on polymer joining involving an interdisciplinary hybrid mechanism.

Dr. Akhil Garg is an Associate Professor at the Ministry of Education's Intelligent Manufacturing Key Laboratory, Shantou University, China. He has been working on sustainable manufacturing processes and optimization methods since 2011. He received his doctoral degree from Nanyang Technological University (NTU), Singapore, in 2014. He has published over 50 SCI-indexed articles in the areas of manufacturing and optimization.

Liang Gao received his B.Sc. in mechatronic engineering from Xidian University, Xi'an, China, in 1996 and Ph.D. in mechatronic engineering from the Huazhong University of Science and Technology (HUST), Wuhan, China, in 2002. He is a Professor in the Department of Industrial and Manufacturing System Engineering, School of Mechanical Science and Engineering, HUST, and the Vice Director of State Key Laboratory of Digital Manufacturing Equipment. He had published over 190 refereed papers. His current research interest includes optimization in design and manufacturing. He currently serves as the Editor in Chief of IET Collaborative Intelligent Manufacturing, Associate Editor of Swarm and Evolutionary Computation and Journal of Industrial and Production Engineering, and Editorial Board Member of Operations Research Perspectives.

Chapter 1
Introduction to Polymer Science

1.1 Polymer

Polymers are chemical compounds containing large molecules joined together to repeat the same type of chains. Polymers are generally termed as plastics in industries. Polymer structure is illustrated basically through a chain. The chain is made of several junctions linked together. In similarity, the atoms inside the polymers are linked to each other in the polymer chain. The molecular linkages are repeat units in the polymer chain created by molecules known as monomers. The repeat unit structure can differ broadly based on the raw materials that form the polymer. For example, in polyvinylchloride, the polymer utilized create a broad range of plastic pipes and electric cables that has a very simple repeat unit that is two carbons that are joined to one another leading to a single linkage.

Polymers find utility in versatile applications. The utility extends to coatings, adhesives, foams, composites, electronic devices, and packaging materials to textile and industrial fibers, optical devices, biomedical devices and precursors for various newly established high-tech ceramics.

Depending on the preferred application, polymers can be slightly altered to leverage the beneficial properties. These properties exhibited involve reflective, tough, impact resistant, translucent, brittle, elastic, inelastic, insulative, soft and malleable features.

1.2 Classification of the Polymers

Various new polymers are synthesized and are further developed in the futurity. Suitably, polymers are segregated into four major groups depending on their structure, origin of source, molecular force and mode of polymerization (Fig. 1.1). Among this classification, polymer structure type is more significant in the polymer science.

S. A. Vendan et al., *Confluence of Multidisciplinary Sciences for Polymer Joining*,
https://doi.org/10.1007/978-981-13-0626-6_1

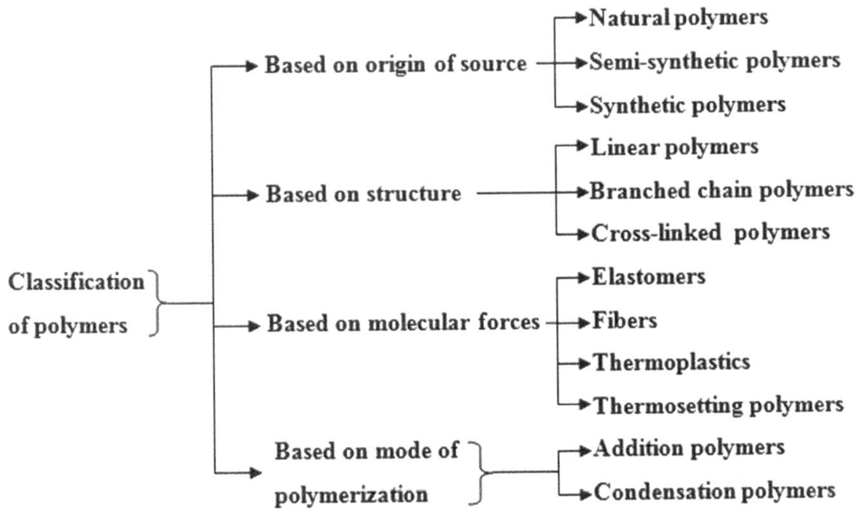

Fig. 1.1 Classification of polymers [1]

The categorization is useful for the reasons that facilitate the characterization of properties. Tacticity, crystallinity and backbone atom of the polymers are other classifications of the polymers.

1.2.1 Based on Origin of Source

Natural polymers: The plants and animals are utilized as the source for producing polymers. For example, natural polymers such as cellulose, starch, proteins, rubber, silk, wool and etc., are produced from the plants and animals.
Synthetic polymers: The synthetic polymers are produced from low weight molecular compounds in the laboratories and industries which are also known as human-made polymers. Examples for synthetic polymers are nylon, polyethylene, terylene, polystyrene, synthetic rubber, polyvinyl chloride, Teflon, and bakelite etc.
Semi-synthetic polymers: Semi-synthetic polymers are derived from natural polymers with enhanced physical properties such as tensile strength and lustrous nature by chemical treatment. Cellulose acetate (rayon) and cellulose nitrate are few examples for the semi-synthetic polymers.

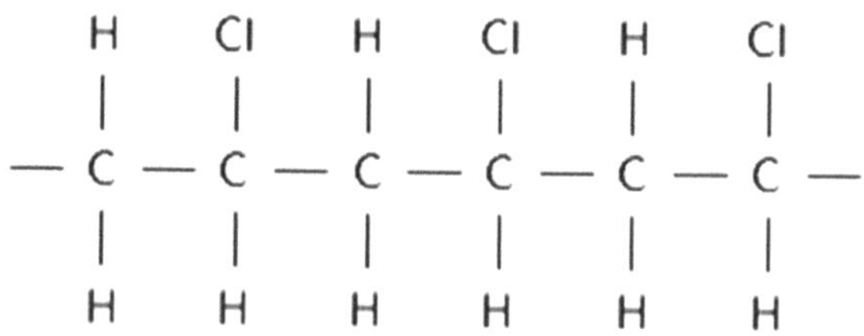

Fig. 1.2 Polyvinyl chloride (linear polymer) [2]

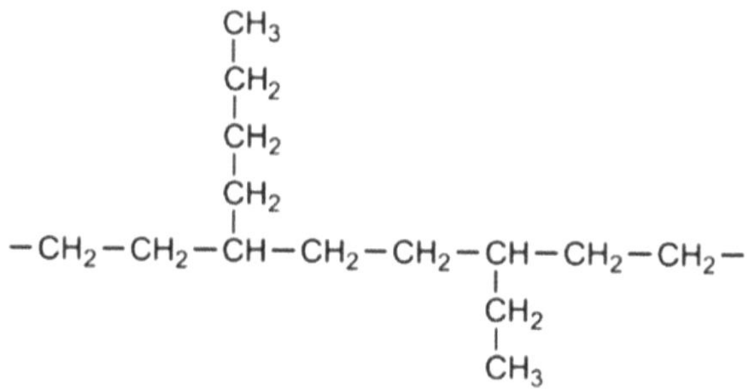

Fig. 1.3 Low density polyethylene (branched polymer) [3]

1.2.2 Based on Structure

Linear polymers: Monomers are linked by long and straight chain without any side chain in this structure of polymer. Polyvinyl chloride (Fig. 1.2) and nylon are some of the examples for linear polymers.
Branch chain polymers: Linear chains of polymers with many side chains are called as the branch chain polymers. Few examples for the branch chain polymers are low density polyethylene and linear low density polyethylene (Fig. 1.3).
Cross-linked polymers: Bi-functional and tri-functional monomers with covalent bonds or ionic bonds are known as cross-linked polymers. Bakelite, formaldehyde resins, melamine and vulcanized rubber (Fig. 1.4) are some examples for the cross-linked polymers.

1.2.3 Based on Molecular Forces

Polymers are also categorized based on various intermolecular forces (Fig. 1.5) which are presented in this section.

Elastomers: Elastomeric polymers are stretched by applying a tensile force. Subsequently, it will return to original geometry when the force is released. The weakest molecular forces (Vander waal forces) are the structures observed in the polymer

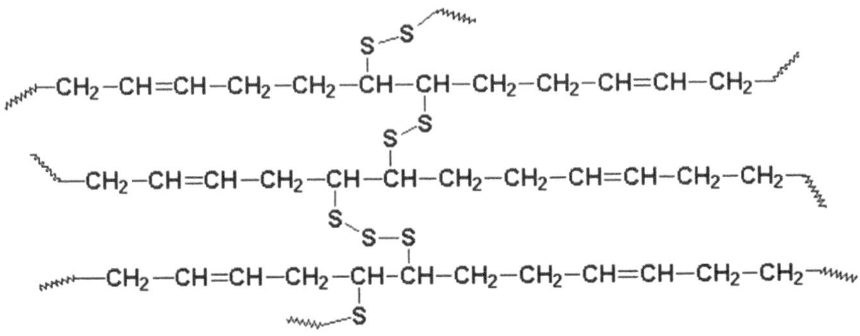

Fig. 1.4 Vulcanized rubber (cross-linked polymer) [4]

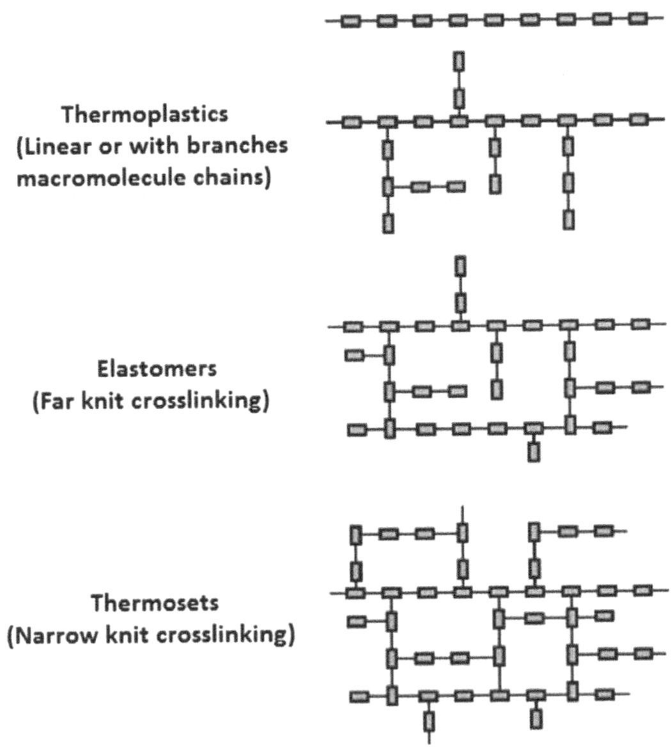

Fig. 1.5 Structures of thermoplastics, elastomers and thermosets [5]

chains of the Elastomers. Few examples for the elastomers are neoprene, vulcanized rubber, buna-s and buna-n etc.

Fibers: Structure of fibers show strong intermolecular forces such as hydrogen bonding. High tensile strength and high modulus are the properties possessed in the fibers due to their structure. Nylon 66, terylene and dacron are few examples for the fibers.

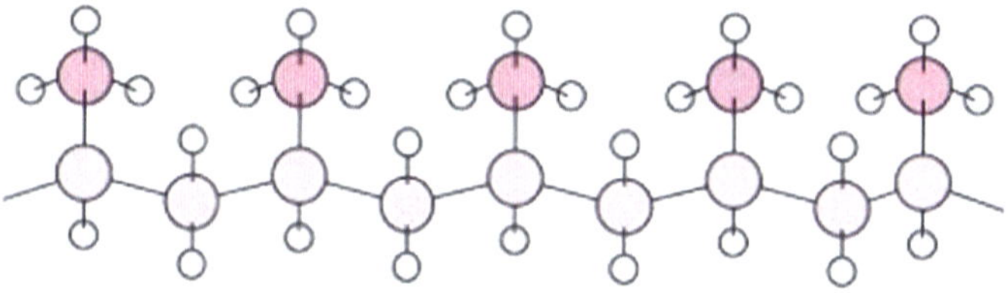

Fig. 1.6 Isotactic polymer (polypropene) [6]

Thermoplastics: Polymers possessing the intermediate intermolecular forces among the elastomers and fibers are known as thermoplastics. Thermoplastics are linear polymers and there is no cross-link chain in the structure. Some instances for thermoplastics are polyethylene, polystyrene and polyvinyl chlorides etc.
Thermoset polymers: These polymers are created from low molecular mass semi-fluid polymers. Thermosetting polymers are extensively branched chain molecules with broad cross-links when heated. The polymer becomes infusible and cannot be recycled. Melamine and bakelite are few examples for thermosetting polymers.

1.2.4 Based on Mode of Polymerization

Addition polymers: Addition polymers (chain-growth polymers) are formed on a large scale by adding repeated units of similar or dissimilar monomers without removal of any by-product. Some examples for the addition polymers are polyvinyl chloride, polyethylene and polypropylene etc.
Condensation polymers: The condensation among two various bi-functional or tri-functional monomeric units which react together with removal of some by-product are known as condensation polymers (step-growth polymers). Nylon 66, bakelite and polyester are some examples for the condensation polymers.

1.2.5 Based on Tacticity

Spatial/geometric configuration of molecules in polymer chain can be arranged into systematic or random configurations known as tacticity. Isotactic polymers (Fig. 1.6) refer to the functional groups that are arranged on the same side of the polymer chain from the head to tail configuration. In the case where, the deposition of side groups are in arbitrary order around the main chain then it is known as atactic polymers (Fig. 1.7). While, the deposition of side groups in varying mode, is termed as syndiotactic polymers (Fig. 1.8). The configurations of the polypropene shown in the figure illustrate various tacticity.

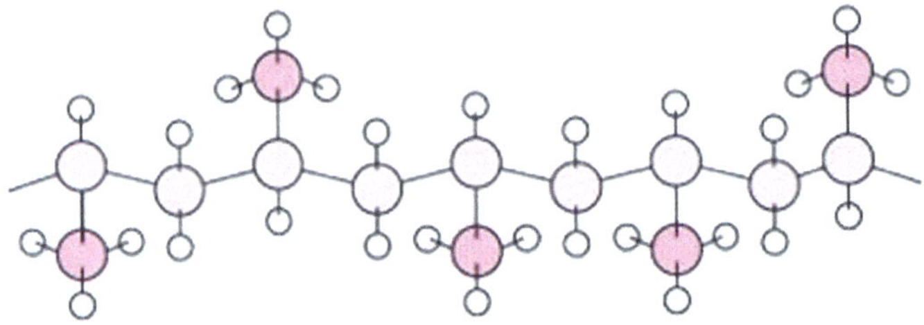

Fig. 1.7 Atactic polymer (polypropene) [6]

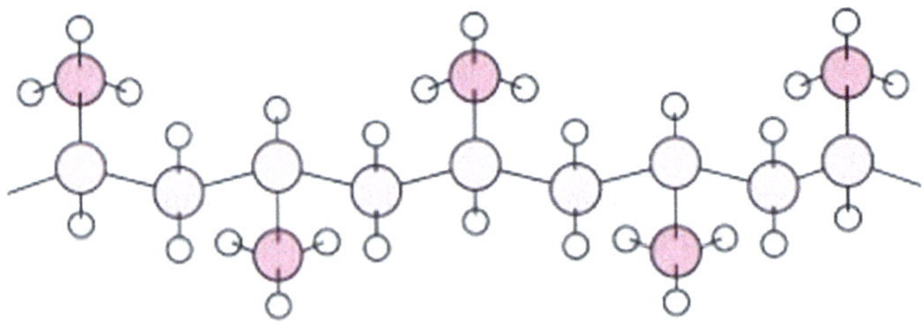

Fig. 1.8 Syndiotactic polymer (polypropene) [6]

1.2.6 Based on Crystallinity

An orderly atomic arrangement is produced by packing of the molecular chains known as crystallinity of polymers. The classified polymers are based on the crystallinity, amorphous and semi-crystalline polymers. Random arrangement of molecular chains of polymer is known as amorphous polymers. Partly systematic arrangement of the molecular chains is known as semi-crystalline polymers.

1.2.7 Based on Backbone Atom

Polymers are also classified based on the chemical structure of their backbone. The major classifications are organic and inorganic polymers. Many repeating monomer units formed by macromolecules are known as organic polymers and it uses carbon as the backbone structure. The macromolecules comprises of many repeating monomer units called as inorganic polymers that comprise of lack of the carbon backbone.

1.3 Engineering Plastics

Engineering plastics are the group of polymer materials that have exceptional mechanical properties such as toughness, stiffness and low creep that are used in

Table 1.1 Comparison of the properties for metals, ceramics and plastics

Property (approximate values)	Metals	Ceramics	Plastics
Density (g/cm^3)	2–22 (average ~ 8)	2–19 (average ~ 4)	1–2
Melting points	Low (Ga = 29.78 °C, or 85.6 °F) to high (W = 3410 °C, or 6170 °F)	High (up to 4000 °C, or 7230 °F)	Low
Hardness	Medium	High	Low
Machinability	Good	Poor	Good
Tensile strength, MPa (ksi)	Up to 2500 (360)	Up to 400 (58)	Up to 140 (20)
Compressive strength, MPa (ksi)	Up to 2500 (360)	Up to 5000 (725)	Up to 350 (50)
Young's modulus, GPa (10^6 psi)	15–400 (2–58)	150–450 (22–65)	0.001–10 (0.00015–1.45)
High-temperature creep resistance	Poor to medium	Excellent	–
Thermal expansion	Medium to high	Low to medium	Very high
Thermal conductivity	Medium to high	Medium, but often decreases rapidly with temperature	Very low
Thermal shock resistance	Good	Generally poor	–
Electrical characteristics	Conductors	Insulators	Insulators
Chemical resistance	Low to medium	Excellent	Good
Oxidation resistance	Generally poor	Oxides excellent; SiC and Si_3N_4 good	–

the production of structural products [7] like bearings, gears, electronic equipments, and auto components. General properties of plastics are compared with metals and ceramics as shown in Table 1.1. They are made up of minimum quantities due to their higher cost and lean to be utilized for smaller components or low-volume usages (such as mechanical components), rather than for high-volume ends and bulk size (such as packaging and containers). The expression generally refers to thermoplastics rather than thermosetting polymers. For example, engineering plastics contain polycarbonates, utilized in motorcycle helmets; acrylonitrile butadiene styrene (ABS), utilized for car dashboard trim, bumpers and Lego bricks; and polyamides (nylons), utilized for skis and ski boots.

1.4 Applications

Engineering plastics are finding variety of applications in the construction sector. Several beneficial factors of the plastics are higher adaptability and combine tremendous strength to weight ratio, low maintenance, corrosion resistance, durability and cost effectiveness. Plastics are utilized in the electrical sectors as the electrical insulation, heat insulation (by adding special flame retardant additives) etc. [8]. Plastics are the appropriate material for packaging goods. They are utilized in abundant packaging applications comprising of bottles, trays, drums, cups and vending packaging, boxes, baby products, protection packaging and containers. Plastics are used in the transportation sector due to their lightweight and lower cost. Reducing the weight of automobiles can lessen fuel consumption severely. In the automobile industries, plastics are used as dashboards, seats and faces, flooring while maintaining their striking exterior and clean trouble-free system.

1.5 Welding of Engineering Plastics

Welding of polymers is through heating to facilitate softening and fusion at the interface. Weldability of plastics is identified based on the molecular forces of the polymers such as thermoplastics, thermosets or elastomers. Particularly, the thermosets are chemically reactive during processing and curing, hence, they produce irreversible cross-linking reactions in the plastics mold. The chemically cross-linked elastomers or molded thermoset parts that can be altered by heat while undergoing degradation. Thermosets and elastomers are only joined by mechanical fastening and adhesive bonding techniques. Conversely, thermoplastics and thermoplastic elastomers can be softened and fused due to the deterioration of hydrogen bonding forces and secondary van der walls along the nearby polymer chains. Thus, thermoplastics and thermoplastic elastomers are recycled by providing sufficient heat to make the mold and welding of the specimens effective.

The fabrication of a successful bond is based on four factors by: the surface free energy, the chemical nature of the polymers, the surface topography, and contamination of the polymer surface by oil, dust, or grease [9]. Aforesaid factors considerably influence the efficiency of the adhesive and solvent joining techniques. Still, fusion welding is much more tolerant to factors like material differences from specimen to specimen and surface contamination. Fusion welding or solvent is used to join for the polymers with the characteristics of solubility.

The appropriate heating of the polymer or solvent is performed to establish the chain diffusion and enable the essential adhesion to the polymer specimens. Previous reports conclude that thermoplastics and thermoplastic elastomers are mostly joined by the fusion welding techniques. The polymer chains can acquire adequate mobility to inter-diffuse when the melting temperature (Tm) in crystalline polymers and the glass transition temperature (Tg) in amorphous polymers are exceeded. Various techniques are available for welding of thermoplastics, thermoplastic elastomers and thermoplastic composites. External sources such as conduction, convection, and/or radiation processes and internal sources such as molecular friction produced by mechanical motion at the joint are the reasons for creation of thermal energy. During external heating, the heat source is eliminated earlier than the usage of pressure, and longer welding times that are evenhanded by the greater tolerance to differences in material properties. Internal heating techniques are based on the material characteristics. Pressure and heating are given concurrently; shorter welding times commonly involve various stages of joining process.

1.6 Welding Terminologies for Plastics

Adhesive bonding: The ultrasonic system is adapted to trigger an adhesive on a tape thus allowing the welding parts that are not generally distorted by ultrasonics alone.
Anvil: The support and positioning of the work piece which are being welded is called as anvil or welding nest.
Booster: It enhances the amplitude of vibration which is placed nearer to the converter.
Converter (transducer): Transducer converts the high frequency electrical signal to mechanical vibration through the piezoelectric crystals.
Clamping force: The action of pneumatic cylinder is used to exert the force on the specimen being welded.
Distance measurement: It is used to determine pre-weld height and the compression of the specimen after it is joined. The device is attached in the welder head.
Energy: Power output of the product and the time of usage, expressed in watt-seconds or joules.
Horn: It is used to provide the vibration to the work piece to be welded.
Kilohertz (kHz): It is the frequency of the signal (1 kHz referred as 1000 cycles/s).
Kilowatts (kW): It is the systems power rating (1 kW referred as 1000 W).
Mass: Mass is connected to the slide and the pneumatic system. It is the large block above the reed.
Nest: It is also named as anvil which is the tool to position or support the components to be welded.
Pattern roller: The pattern roller cuts or embosses a pattern or sews and aids the film or fabric being joined for rotary textile welding.
Power supply: It is used to deliver the power through plugs and converts the electrical signal to the high frequency, high voltage signal required to push the ultrasonic system.
Plunge welding/cutting: It is the action of cutting or welding occurring in a single pulse and the cut pattern or weld region is verified by the shape and size of the anvil and horn.
Reed: The welding tip is held by the vertical rod and part connected to the wedge is known as the reed in ultrasonic welding.

Repeatability: It is static characteristic to generate results which are necessarily the same when other process parameters are kept constant. It is also known as a strong technique or controlled process.
Seam welding: Ultrasonic energy is given to produce continuous seam welds while a roller is rotated and traversed across the material to be joined.
Sonotrode: It is the welding tip of ultrasonic machine.
Spot welding: Spot welds (circular, elliptical, annular, and rectangular, etc. in geometry) are produced when the material is fixed between the anvil and the shaped tip and subsequently ultrasonic energy is subjected.
Taper lock tip: The taper lock tip is used with the wedge-reed system. It is a single piece welding tip with a Morse taper which suits into a pairing tapered container in the reed. And this offers an inexpensive, exchangeable tip generally made of heat treated steel which can be adjusted before locking into place.
Trigger: The firing of ultrasonic energy is caused by the variable, generally with a preset pressure or time cycle.
Ultrasonic: Signals whose frequency range between 15, 20 kHz or higher and lies above the standard hearing range.
Ultrasonic bonding/welding: Weld or bond of plastics, synthetic textiles or specific metals is caused by the high frequency acoustics vibration with pressure.
Wedge-reed: Wedge is included in the assembly, and reed sends out the ultrasonic acoustic frequency energy to the welding tip and mass block in the course of the force application.
Weldable materials: Thermoplastics sharing compatiblity with the welding of ultrasonics is used in the textiles and plastics industry. Examples of the weldable plastics are polyethylene, polyester, acrylics, polypropylene, nylon, acetate, polycarbonate, ABS etc. Instances of the weldable metals are generally copper and aluminum alloys. Weldable metal configurations can be wire or sheet metal, and tubes of up to 9 mm diameter that can be bonded without brazing and crimping.

References

1. Brinson, H.F., and L.C. Brinson. 2008. Polymerization and classification. *Polymer engineering science and viscoelasticity: An introduction* 99–157.
2. Naqvi, M.K. 1985. Structure and stability of polyvinyl chloride. *Journal of Macromolecular Science-Reviews in Macromolecular Chemistry and Physics* 25 (1): 119–155.
3. Richards, R.B. 1951. Polyethylene-structure, crystallinity and properties. *Journal of Applied Chemistry* 1 (8): 370–376.
4. Flory, P.J. 1944. Network Structure and the elastic properties of vulcanized rubber. *Chemical Reviews* 35 (1): 51–75.
5. Staff, P.D.L. 1997. *Handbook of plastics joining: A practical guide*. William Andrew.
6. van Koten, G., H. Hagen, and J. Boersma. 2002. Homogeneous vanadium-based catalysts for the ziegler-natta polymerization of a-olefins. *Chemical Society Reviews* 31 (6): 357.
7. Lampman, S. 2003. *Characterization and failure analysis of plastics*. ASM International.
8. Klein, R. 2012. *Laser welding of plastics: Materials, processes and industrial applications*. Wiley.
9. Troughton, M.J. 2008. *Handbook of plastics joining: A practical guide*. William Andrew.

Chapter 2
Polymer Welding Techniques and Its Evolution

2.1 Evolution of Polymer Welding

The physics underlying the polymer joining is autohesion. The formation of stable bonds between two surfaces of specimens in contact is known as autohesion process. Restriction to the peeling of the welded parts at the primary contact surface is offered by the bonds formed. Autohesion method is initially reported in the year of 1935–7 [1]. Autohesion mechanism is reported by several researchers. The existence of free molecular ends on the contact surface is associated with the autohesion. Autohesion bonds formed by diffusion at the contact surface with molecular ends are assumed and reported by J. R. Scott (Fig. 2.1). The validation of the diffusional nature of autohesion is performed by direct and indirect procedures. However, polymer welding based on autohesion is not controlled by diffusion as reported from the literature. In the polymer joining, the mechanism and stages of the process influence the strength of welded joints. Hence, evaluation of polymer welding processes is essential for forecasting the mechanical properties of the bonded joints. The mechanism of polymer welding involves the entire combination of the procedures which occur from the initiation to the ending. The factors that play significant role in polymer joining are surface condition of specimens to be welded, temperature and pressure experienced by the specimens during the course of the joining process [1]. Sequences followed in the welding mechanism generally can be categorized into two groups:

1. Procedures which recognize the joining of the parts being welded;
2. Procedures which make conditions for the first sequence to proceed. The first series of operations includes:
    (a) Diffusion of macro-radicals, molecules or molecular segments of the polymer which can either be in solid, liquid, or dissolved state;
    (b) Convective mass transfer;
    (c) Recombination of macro-radicals across the contact surface;
    (d) Physical (surface) interaction;

S. A. Vendan et al., *Confluence of Multidisciplinary Sciences for Polymer Joining*,
https://doi.org/10.1007/978-981-13-0626-6_2

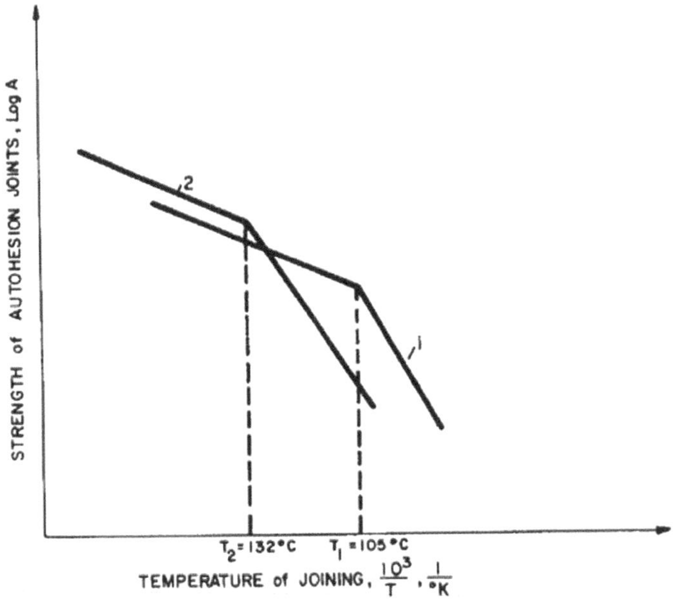

Fig. 2.1 Relation of strength of autohesion joints on joining temperature. (1) Low-density polyethylene (LDPE) and (2) high-density polyethylene (HDPE)

(e) Any combination of processes described above, physical (surface) interaction occurs in every mechanism. If the surface interaction does not comply with the joining, the physical interaction is attributed to the second group which contains:
(a) Formation of the real contact surface;
(b) Formation of the macro-radicals;
(c) Destruction and removal of inert layers which prevent real contact of active material being welded.

Several of the processes mentioned above can be described by the same Arrhenius equation but with different constants. Therefore, if the equation for description of the entire process is known, it does not always mean that the limiting process can be determined. Moreover, the rate of different processes is a function of welding conditions but not the same function. So, at two different temperatures of welding, two different processes can limit the rate for the overall welding operation. Such an exchange is shown on the graph in Fig. 2.9, where both parts of the broken line can be described by the exponential equations, but activation energies in these equations vary by approximately a factor of ten [1].

Autohesion processes occur on the closed surface and spread out to about 10–100 angstroms in depth from the contact surface. So, it is difficult to study the welding mechanism by direct methods. The mechanism of plastic welding can be understood if the rates of different processes that occur during the operation are various physical conditions [1].

2.2 Recent Issues on Thermoplastic Welding

Minimizing the time and improving the quality are the key objectives for any industries. Polymer material suits the requirement, and the mixture of various polymer materials is essential for certain specific designs. Polymers are predominately adopted in a wide range of advanced applications, particularly in optical materials, coatings, semiconductors, aerospace, automotive parts, composites, and medical devices [2]. Metals are gradually replaced by the thermoplastic polymers in the aerospace and automotive industries over the years. The replacement of metals with plastics is due to the reduction in the cost. Significant factors are limited to not only minimizing incurred cost but also weight reduction, enhanced thermal insulation, and flexibility features.

Various innovative designs are proposed and implemented owing to the flexibility of the polymer materials. Consumer products like mobile phones and computers have adopted the polymers which have numerous merits such as electromagnetic signal transmission and electrical insulation [3]. Two different classifications of polymer frequently used in the industries include elastomers and thermoplastics. Thermoplastics are preferred polymers due to its thermal and chemical stability and ease in recyclability. Mechanical properties like high fracture toughness and fatigue resistance are typically observed in the thermoplastics [4]. Thermoplastics will soften and finally melt when it is heated above the glass transition temperature (Tg). After the solidification process, the polymer materials retain its original properties with negligible alterations. Several commonly utilized thermoplastic materials are polypropylene (PP), polyetheretherketone (PEEK), polyethylene (PE), acrylonitrile butadiene styrene (ABS), and poly(methyl methacrylate) (PMMA). Reheating and remolding of the thermosetting polymers are not possible. While reheating, the hardening is permanent and does not soften at elevated temperatures with the occurrence of degradation. This evades the weldability of thermoset polymers. Polymer welding methods are classified based on the heat application modes such as internal or mechanical, external, and electromagnetic heating techniques such as in the case of ultrasonic and vibration welding, hot plate welding, and induction and laser welding, respectively [5–11].

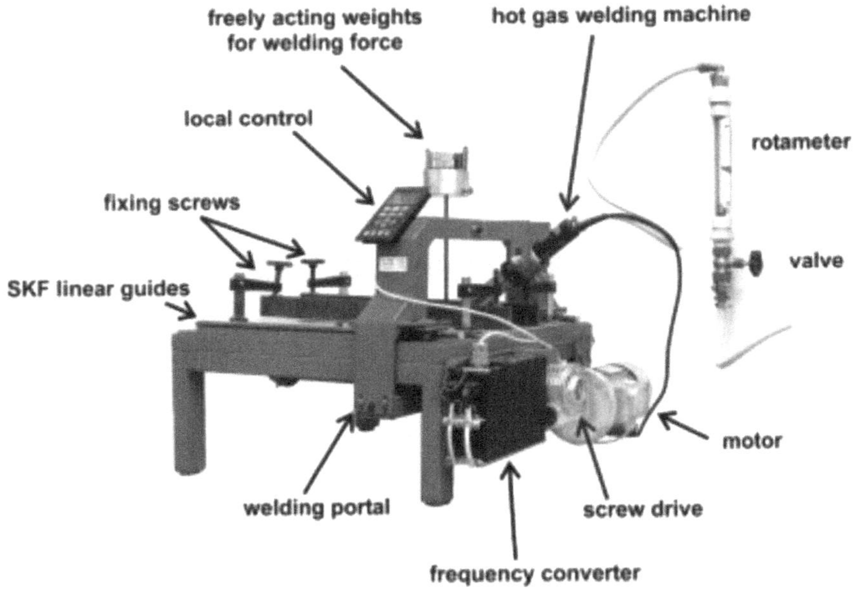

Fig. 2.2 Structure of the automated hot gas welding station [12]

2.3 Hot Gas Welding

2.3.1 Process

Hot gas welding or external heating method uses heated gas to raise the surface temperature of the materials to the softening point [12]. This technique is used for joining and repairing similar materials. Automated hot gas welding station is shown in Fig. 2.2. Joining of fitting plastic basins, fastening small series production, fixing damaged car bumpers, and chemical containers are subjected to the hot gas for compact joining. Air or gas is preheated and sent through the filler and base materials in this technique [13]. The gas is heated above the melting point of the specimens to be welded in this technique [14].

The hot gas stream is used to heat a welding rod and weld groove till they soften to ensure fusing subsequently; the welding rod is compressed into the weld groove. Plastic constructions use hot gas method which is simple, economical, portable, and an apt process for complex design. Schematic representations of hot gas welding are shown in Fig. 2.3. Hot gas technique is employed in sealing of roof/floor membranes, the repair of large injection-molded components, and the fabrication of chemical containers [15].

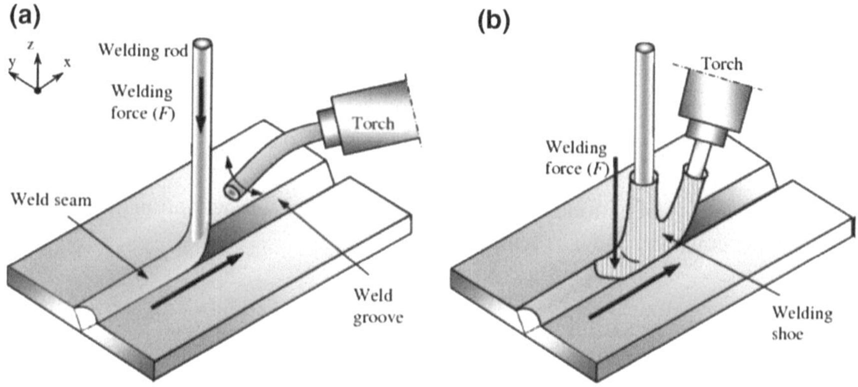

Fig. 2.3 Schematic representations of hot gas welding (HGW). (**a**) Without welding shoe and (**b**) with welding shoe [13]

The materials are heated by hot gas until it reaches glass transition temperature. Generally, nonflammable gases like air, carbon dioxide, nitrogen, and flammable gases such as oxygen and nitrogen are used in this technique [16]. Air is mainly used as gas for this method [17].

2.3.2 *Critical Process Parameters and Phenomena*

Selection of parameters is the key concern to accomplish good joints between the two parts. Welding pressure, welding speed, and gas temperature are the processing parameters governing the bonding of materials [18]. Gas temperature selected depends upon the glass transition temperature of materials or the melting range leading to molecular diffusion of two faces which strike each other at their melt temperatures. The disturbance of the molecules in the materials occurs while heating the polymer material that contributes to disentanglement of the polymer molecules. Aforesaid phenomenon directs the molecules to lose their chain ends. This facilitates sound adhesion between two materials which are in contact as they slightly exceed their melt temperatures.

Flow of molten polymer occurs at the interface between the bonding specimens for all the welding methods. Weld strength will reduce with the reduction in the melt flow and mitigate the diffusion at the interface. On the other hand, rise in the melt flow produces stern thermal deformation in the weld zone. It is observed that the key parameter for good weld quality is the flow behavior. For the molten polymer to flow, the movement of the macromolecular chain must be effortless. And it requires sufficient thermal energy to overcome the energy obstacle to achieve the macromolecular chain movement easily. This energy obstacle is denoted as activation energy (Ea). Arrhenius equation (Eq. 2.1) is used to calculate the activation energy

[19]. It is observed that weldability turns out to be good, while polymers' activation energy is less.

$$\eta = C\exp\left(\frac{Ea}{RT}\right) \tag{2.1}$$

where η is viscosity of a molten polymer, C is constant, Ea is activation energy, R is universal gas constant, and T is temperature.

2.3.3 *Prominent Observations in Hot Gas Welding*

It is evident that the single-layer extrusion welding can act as a substitute method for the multilayer hot gas welding of thick polyethylene (PE) sheets (30 mm) with partly or fully plasticized welding rod [20]. The combination of the extrusion and hot gas welding yielded the best results in the strength tests. Polyethylene (PE), polypropylene (PP), polyvinyl chloride (PVC), ABS, FEP, PVDF, PFA, PMMA, and HDPE are the materials generally preferred. This technique is rapid, and it eliminates several problems encountered in other joining method [21]. Various weld factors, materials, and remarks are recorded in Table 2.1.

With the elongations, the welded samples tend to crack at welds. Yield stresses of base materials are higher than the welded samples (Fig. 2.5). The locally expanded structure inside the weld contributes to morphological change in the crystallization state based on the thermal history [22].

The viscosity of the molten material on the bonding surface is reduced by the high welding energy, and this enhances the weld strength using diffusion. Ratio of weld width to weld heat input lessened with an increase in the weld strength [23]. Heat-affected zone (HAZ) is broadened with an increase in the weld energy. Welding speed, gas temperature, and pressure were the factors for determining the bond quality of hot gas welds [24]. The comparison was made for the hot gas butt welding with various materials such as PE, PP, and PVC [25]. The overall tensile strength of the PE-PE and PP-PP bonds was found to be higher than the equivalent PVC-PVC bond. PVC bonds differed in the variety of 45–77% of the base material strength evaluated to 77–90% and 63–80% for PE and PP, respectively. The weaker PVC bond strength was attributed to the higher degree of amorphous characteristic of the PVC, where PE and PP had superior degree of amorphous behavior of the PVC and where PE and PP had superior crystalline content (Fig. 2.4). From the study, the strength of the base material was not attained from this welding technique and welds failed in the weld line. Researchers made a comparison of weld strength of various thermoplastics, where amorphous polymers offer lower weld strengths as a result of a slothful melt flow that directed to chemical putrefaction and consequently degradation occurs swiftly [26]. For high tolerance and high volume production, hot gas and extrusion welding (manual technique) are inappropriate which necessitates

Table 2.1 Various analyses of hot gas welding

References	Types of materials used	Observations
Sims et al. [38]	Polypropylene (PP), polyamide 6 (PA6), PVC, PMMA, and PC	Hot gas welding fabricated a simpler combination of materials in lesser amount without detectable certain content
Marczis and Czigány [3]	Polypropylene	The evaluated welding strength values are compared with welding energy which is obtained from the welding parameters and welding force that can be reported with a quadratic surface function
Atkinson and Turner [39]	Polycarbonate/polyester blend, poly(butylene terephthalate), and an ethylene-propylene-diene rubber	Porosity is lessened if the hot gas can get away from below the joint
Greenawalt [41]	Thermoplastics	The two heat sources at opposite sides of the sealing regions allow an enhanced rate of sealing and an enhanced cyclic rate of the equipment
Haque et al. [40]	PVC	PVC's melting temperature is not defined due to the huge dispersion in crystalline particle size which yields a wide melting range
Toss [42]	Thermoplastics	Temperature variations took place that is complex to realize, and the outcome is moderately high power consumption
Balkan et al. [4]	Polyethylene (PE), polypropylene (PP), and polyvinyl chloride (PVC) sheets	The best outcome was achieved for the double V-joints with welding shoe

skilled labor. Major problems encountered in the hot gas welding and its solutions are given in Table 2.2 (Fig. 2.5).

2.4 Hot Tool Welding

Hot tool welding is the fundamental technique adopted to weld two plastic parts. Hot metal surface is adopted to join the two faces of the specimen with direct contact [27]. The direct contact melts the surface followed by pressing action to form the weld.

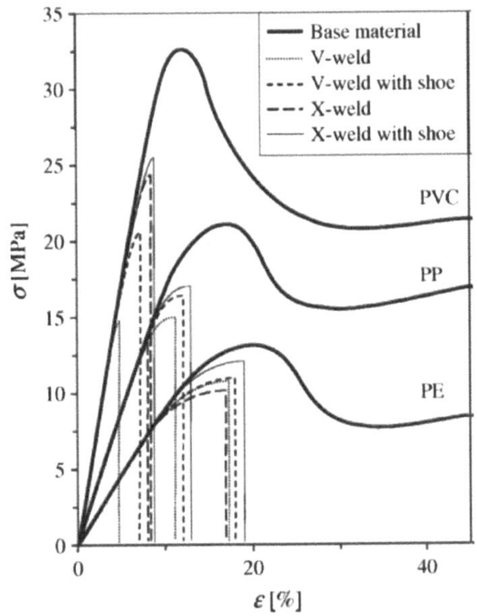

Fig. 2.4 Stress–strain curves for base and welded PE, PP, and PVC sheets with various welding processes [26]

Fig. 2.5 SEM micrograph of double V-welded and fractured PVC sheet [26]

This method varies from heat sealing which packs the material together followed by application of heat to join the parts. Heat sealing is restricted to join the thin films, whereas thick plates are welded by the hot plate welding. Four phases of the hot tool welding are illustrated through the pressure-time diagram as shown in Fig. 2.7 [28]. Heating element used in the hot tool welding is shown in Fig. 2.6.

The components are subjected to a hot tool with high pressure which ensures complete contact between surfaces in phase 1. The parts held together are pressurized until the flow of the molten material starts laterally. In the phase 2, the melt pressure reduces to permit the molten part to solidify. Heat conduction in the molten

Table 2.2 Prominently encountered issues and remedies

Welding method	Prominently encountered issues	Remedy
Hot gas welding	Joint strength is lesser than the parent material	Appropriate jigs, fixtures, and roller will be used to boost the welding pressure applied on welding rod for proper fusion
	High temperature of gas and oxidation produced thermal degradation weakening the joint	Oxidation may be eliminated by shielding the weld region with nitrogen and other inert gases

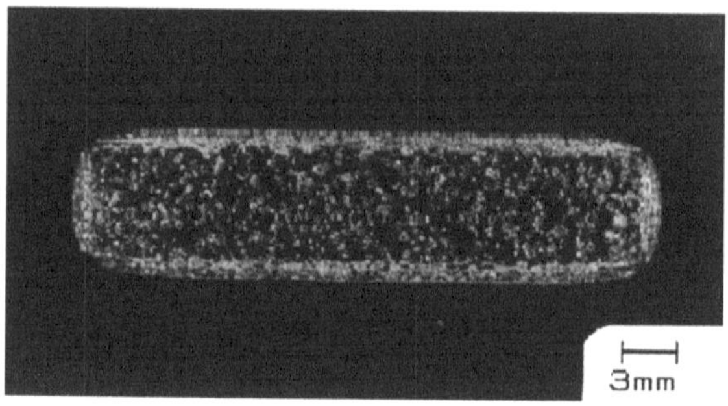

Fig. 2.6 Heating element used in the hot tool welding [18]

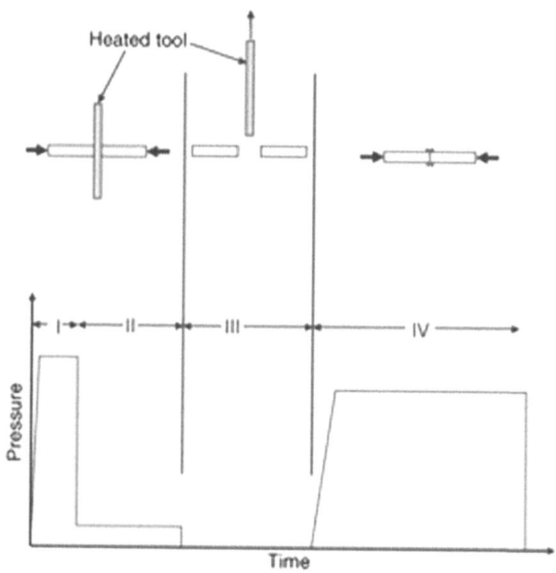

Fig. 2.7 Pressure–time diagram of hot tool welding [18]

layer is controlled as the rate of film thickens. After molten film thickens adequately, the hot tool and parts are separated. To avoid premature cooling of the molten film, the duration reduced in the third phase which is identified as changeover phase. The molten interfaces of the specimens to be welded are subsequently held together and subjected to pressure until the solidification of the weld. While undergoing solidification, molten plastics flow laterally outward. The significant welding factors for this method are the matching pressure. During phase 1 and 2 melting pressure plays a vital role. As the process proceeds in these phases, the temperature of the hot tool is the significant governing parameter. Finally, the re-joining and separation velocities combined with the welding pressure majorly influences the phase 3. The results emphasize that the impact on the process conditions in the molten layer is based on the temperature sensitivity [29]. In this hot tool welding method, the dissimilar materials are joined using various hot tool temperatures for the two components of an assembly [3, 30].

2.4.1 *Thermal Expansion Effect*

Specimen expands during hot tool welding process. The material face is exposed to the hot tool whose temperature varies with time due to lateral flow of the molten material. The equation for the temperature distribution in the specimen is expressed as in Eq. (2.2), and thermal diffusivity '**k**' is assumed as constant [18].

$$T(x,t) - T_a = (T_H - T_a)erfc(x \big/ \sqrt{4kt}) \tag{2.2}$$

where T_a is initial temperature of the specimens, T_H is hot tool temperature, *erfc* is complementary error function, and t is time taken for the period of joining.

Thermal strain is given by Eq. (2.3)

$$\varepsilon_T(x,t) = \alpha(T_H - T_a)erfc(x \big/ \sqrt{4kt}) \tag{2.3}$$

Integrate the $\frac{du}{dx} = \varepsilon_T(x,t)$, where '**u**' is the displacement.

Determination of induced temperature is increased in the specimen length δT as shown in the equation (Eq. 2.4) below

$$\delta T = \int_0^\infty \alpha(T_H - T_a)erfc(x \big/ \sqrt{4kt})dx \tag{2.4}$$

$$= \alpha(T_H - T_a)\sqrt{4kt}\int_0^\infty erfc(\xi)d\xi$$

$$= \alpha(T_H - T_a)\sqrt{4kt}.ierfc(0) = \frac{2\alpha}{\sqrt{\pi}}(T_H - T_a)\sqrt{kt} \tag{2.5}$$

Thermal diffusivity and thermal expansion coefficient are independent of temperature which is assumed constant. This constant estimation is a representative of values for 'k' and 'α'. Thermal diffusivity of polycarbonate (PC) varies from viscosity of 0.172 $mm^2\ s^{-1}$ at an initial temperature of 20 °C to viscosity of 0.099 $mm^2\ s^{-1}$ at a maximum temperature of 150 °C. The expansion coefficient of PC decreases continuously from $6.4 \times 10^{-5}(°C)^{-1}$ at 20 °C to $6.2 \times 10^{-5}(°C)^{-1}$ at 140 °C. It undergoes a large variation near the glass transition temperature, Tg = 155 °C above Tg; α varies from $2.2 \times 10^{-4}(°C)^{-1}$ at 155 °C to $2.0 \times 10^{-4}(°C)^{-1}$ at 275 °C and from $1.9 \times 10^{-4}(°C)^{-1}$ at 350 °C to $1.9 \times 10^{-4}(°C)^{-1}$ at 410 °C.

If air gets trapped between surfaces during melting, it tends to weaken the weld. Weld strength is not significantly affected by few bubbles at weld zone, but will create complexities as the number of bubbles increases. To attain the high weld strength, hot tool was heated up to the required temperature. Weld strength achieved was around 50% by using glass-filled material with 3-mm thickness over a broad range of temperature (T_H = 230–380 °C). The heating time and outflow of the molten material are the parameters used to control the thickness of the molten film during final joining phase [31]. Hot tool temperature range of T_H = 230–260 °C with heating time of t_H = 10 s is the factor to attain the relative weld strength (100%) with strain to failure (21%) in ABS. With temperature range of T_H = 290 °C, PC/ABS material has the maximum relative weld strength (91%) with a failure strain (2.6%). With various temperature ranges T_H = 320 °C and T_H = 290 °C for a heating time t_H = 10 s and t_H = 15 s, PC/PBT material has the very high relative weld strength (98%) with failure strain (4.7%). With various temperature ranges of T_H = 305 °C and T_H = 290 °C and heating time of t_H = 10 s and t_H = 15 s, M-PPO material has the relative weld strength (80%) with failure strain (2%). With temperature range T_H = 305–365 °C and heating time t_H = 10 s, PPO/PA material has the relative weld strength (73%) with failure strain (3.2%). With temperature range of T_H = 305–365 °C and heating time of t_H = 10 s, 30 GF-MPPO materials have the relative weld strength (71–77%) with failure strain (2.2%). With temperature range of T_H = 275–305 °C, PC/ABS to M/PPO materials have the relative weld strength (30%) with failure strains of 0.6% [32].

Besides, the surface roughness of rapid fracturing detailing (RFD) component might be enhanced due to the melting of polymer surface that fills in pores at the stage of operation. Specially, as a result of the non-contact tool behavior, the amount of energy received by the specimen is based on thermal radiation across the gap among tool and specimen. So, the equation (Eq. 2.6) is defined for the effective heat input per unit of gap combined by the gap factor [33].

$$Q_{eff}^{h} = \frac{Q_{eff}}{H} = \frac{Qi}{V_C H} \tag{2.6}$$

2.4.2 Critical Process Parameters and Phenomena

The most significant process parameters for hot tool welding are the dwell/changeover time, the temperature of the hot plate, the heat soak time, the welding pressure, the heat soak pressure, the welding time, the joining displacement, and the cooling time. From the various process parameters, geometry of weld interface, heating time, and hot plate temperature are essential parameters to govern the properties of hot plate welding of plastics. Weld strength is strongly influenced by hot plate temperature, melt flow, and heating time at the duration of welding of poly (methyl methacrylate), glass fiber-reinforced polypropylene, and polypropylene bars which is revealed from the investigation [34]. Polypropylene weld strength is majorly affected by process parameters of hot plate welding such as heating time and hot plate temperature. It is observed from the optimal weld strength of the polypropylene using design of experiments [35]. In this investigation, various geometries like flat, semicircular, rectangular, and triangular weld interfaces are the parameters affecting the weld strength as observed for polypropylene with and without glass fiber-reinforced composites. Heating time and hot plate temperatures are used with the geometrical parameters in this study which affects the joint properties [36]. Welding parameters (plate temperature, heating time, and welding displacement) of hot plate welding affected the weld strength as per the tensile test results of polycarbonate (PC)/acrylonitrile butadiene styrene (ABS) blends. The joint strength is enhanced about 70% regarding parent material properties using plate temperature of 290 °C, welding displacement of 1.25 mm, and heating time of 25 s [37].

2.4.3 Prominent Observations in Hot Tool Welding

Squeezing the entire melt out of the bond line is prevented by mechanical stops which yields cold weld. Good heat conduction and little pressure among hot plate and joining components may be uncertain. The applied pressure should be sustained till the thermoplastic matrix starts to soften and flow [38]. Hot tool welding is influenced by the mechanical and thermal properties of the pipe materials. Restructuring of crystalline stage occurs, while the crystalline zone with greater mechanical and thermal properties is exhibited due to larger crystallite size [3]. To achieve described shapes on surface of VLM-ST component, rapid fracturing detailing (RFD) method using the non-contact heat tool was established. The heat of the tool creates localized melting between the surfaces of specimen is the principle of RFD method. Major issues encountered in the hot tool welding and its solutions are given in Table 2.3.

Table 2.3 Prominently encountered issues and remedies

Welding method	Prominently encountered issues	Remedy
Hot tool welding	Insufficient melt dwell time	Increasing melt dwell time improves penetration with the plastics facilitating sufficient melting of the parts
	Higher changeover time	Higher changeover time converts the plastics semimolten stage creating skin on the surface of the workpiece. In order to overcome the change, reduction of the tool retraction time and sealing the workpieces are essential
	Insufficient seal force	Insufficient seal force weakening the weld due to improper mix of molten plastics. Provision for adequate sealing force will lead to proper mixing of molten plastics

2.4.4 *Resistance Welding*

Resistance welding is used in industries for bonding metal sheets and parts. The metal components are melted due to heat generated by current conduction. Melting occurs at the localized points which are predetermined using the design of the electrodes and/or the specimens to be joined. A calculated force is provided from the initiation to the end of the process while applying current to limit the contact region at the bond interfaces and, in some usages, to forge the specimens. Resistance welding presents a number of merits and usually eliminates surface treatment [39, 40]. Resistance welding is also referred to as resistive implant welding [41]. Energy generated by resistance welding is computed from the equation of $E = I^2Rt$ which is Joule's law. When current discharges through the heating part, then the energy (E) created by the resistor is proportional to the resistance (R), elapsed time (t), and current (I).

The setup for resistance welding of single-lap process is illustrated in Fig. 2.8. The process of the resistance welding is based on the parameters such as geometry, heat transfer, and temperature-dependent material properties. While the given energy surpasses the thermal losses, the temperature of the sheets begins to increase, first in the nearer purview to the joint surfaces and afterward deeper into the material. While the temperature at the joint boundary increases to a specific point, thermoplastic matrix begins to melt. While minimal melting is attained with precise time, the discharge of current is stopped and the bond is permitted to solidify when required pressure is maintained. Plastic resistance welding is different from metallic resistance welding, wherein the welding pressure is maintained throughout the process which creates the close contact among the sheets and the interlayer. The heating component is used to join the specimens at the time of the joint production method [42].

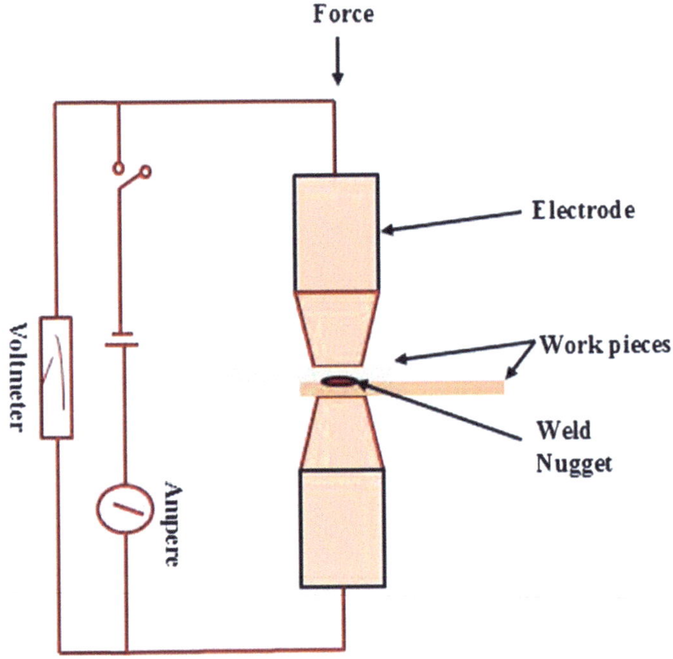

Fig. 2.8 Resistance welding process [40]

2.4.5 Critical Process Parameters and Phenomena

The significant factors for the welding method are the input energy E in kJ/m^2 (Eq. 2.8) and the input power P in kW/m^2 (Eq. 2.7), which are given at the bond interface per unit area:

$$P = I^2R/LW \tag{2.7}$$

$$E = Pt \tag{2.8}$$

where R is the resistance of the heating component of the welding region (in Ω), I the current provided to the heating component (in A), E the thermal energy (in kJ/m^2), P the power (in kW/m^2), t the electrified time (in s), and W and L the width and length (both in m) of the welding region [4]. From the results of several trial and error tests, the nominal agreeable power level was obtained which is 80 kW/m^2. Minimum melting time can be maintained for high power levels due to the greater heating rate. Due to the heating up of the matrix and fiber oxidation, bond efficiency lessens while offering high power level (above 200 kW/m^2). In fact, various tests were conducted among the two power limits which are from 80 to 160 kW/m^2.

Various power levels are adopted for the lap shear strengths of workpieces bonded as illustrated in Fig. 2.9 as a function of thermal energy released by the heating com-

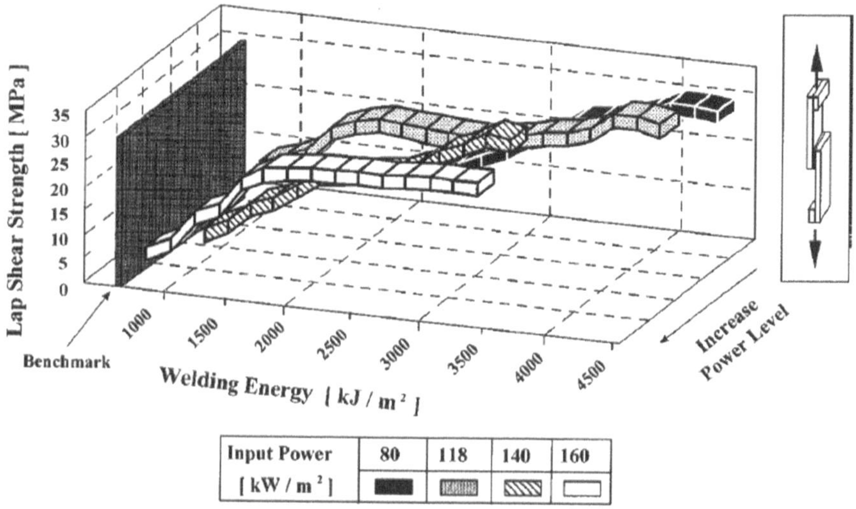

Fig. 2.9 Effects of welding energy and power level on lap shear strength [4]

ponent. The benchmark lap shear strength of a compression-molded workpiece is represented as the black column values. Two trends are determined from Fig. 2.9 which is the energy required for the peak value of lap shear strength that diminishes considerably with raising power level and the lap shear strength of the bonded workpieces that moves toward a peak level with raising energy which gradually lessens little for each power level.

2.4.6 *Prominent Observation in the Resistance Welding*

Post-failure surface is observed with raising welding energy and due to change in the failure approach from interfacial to intra-laminar failure [43]. Fracture surface of the bonded workpiece for conservative and beneath three-point bending is illustrated in Fig. 2.10.

An electrical resistance drop was recorded followed by the melting of the polymer at the weld interface as a result of the afresh electrical current paths made in the adherents. Afresh-insulated heating component comprising of a TiO_2-coated stainless steel mesh effectively eliminated current leakage. This new method eliminated electrical resistance drop. The temperature uniformity was enhanced by the TiO_2 heating component and lessened the welding time by avoiding power losses in the adherents. The drawback of the TiO_2 heating component is its inability to improve mechanical performance [44]. After a polymer mechanical interlocking rupture, the failure stage of the UD workpiece depended on ceramic debonding. The failure stage was also seen before applying the static loading. Various analyses performed by researchers

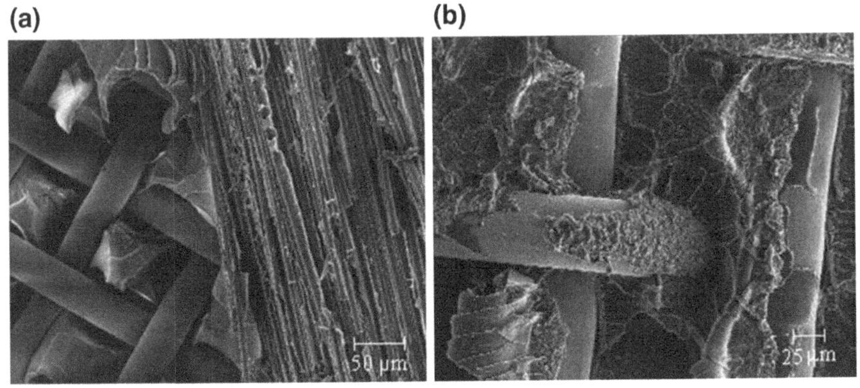

Fig. 2.10 Fracture surfaces of quasi-isotropic specimens welded using the conventional. (**a**) TiO_2 and (**b**) HE, under three-point bending [44]

in resistance welding are given in Table 2.4. Major problems encountered in the resistance welding and its solutions are given in Table 2.5.

2.4.7 *Laser Welding*

In laser welding, two or more parts of material are welded together through application of a laser. It is a non-contact technique that needs access to the bonding region from one side of parts being joined as shown in Fig. 2.11. The joint produced by the depth laser light swiftly heats the material in milliseconds. The laser beam has high energy content and low beam divergence which generate heat, while it buffets a surface.

The major categories of lasers utilized in welding and cutting are:

- Gas lasers use a combination of gases like helium and nitrogen cascaded to carbon dioxide laser systems. These lasers use a high-voltage power source, low current to electrify the gas combination by a laser medium. These lasers work in a continuous or pulsed mode. Carbon dioxide lasers utilize the combination of pure carbon dioxide with nitrogen and helium as the laser medium. CO_2 lasers are also used in dual-beam laser joining where the beam is divided into two equivalent power beams.
- Solid-state lasers: It works at 1 μm wavelengths. They can be pulsed or work incessantly. Pulsed operation created bonds identical to spot welds but with greater depth of penetration. The parameters for solid-state lasers like Nd:YAG type and ruby lasers are pulse time (1–10 ms) and pulse energy (1–100 J).
- Diode lasers.

Table 2.4 Different analyses performed in resistance welding

Author(s)	Material	Heating element	Analysis
Hou et al. (1999)[43]	CF/ PEI	CF/PEI (semiconducting nature of carbon fiber) Advantages of heating element: 1. Minimizes the negative effects during welding time 2. Smooth contact surface	Nondestructive evaluation technique and destructive tests are performed. The relationship between input power, input energy, welding time, and pressure are obtained in this study
Dube et al. (2007) [56]	1. UD APC-2/AS4 2. UD GF/PEEK 3. Quasi-isotropic APC-2/AS4	1. Stainless steel mesh (conventional) 2. Two films 3. TiO_2 4. Stainless steel mesh 6. TiO_2	Mechanical testing is carried out using short beam and three-point testing. TiO_2-coated stainless steel mesh successfully prevents current leakages during welding time (melting phase)
Shi et al. (2013) [36]	UD APC-2/AS4	Stainless steel	Fatigue failure characterization is obtained from the experiments using three-point bending which showed the better bonding of the specimens
Jakobsen et al. (1989) [63]	PEEK APC-2 (different thicknesses)	APC-2	One-dimensional transient heat transfer analysis using finite difference method is carried out to obtain temperature profile. Heat transfer analysis coupled with degradation kinetics model is performed to determine the thermal degradation of the specimen
Pitchumani et al. (1996) [64]	CF/PEI GF/PEI	Stainless steel	Fracture surface analysis revealed the better bond strength in the specimens

(continued)

Table 2.4 (continued)

Author(s)	Material	Heating element	Analysis
Xiao et al. (1992) [34]	Quasi-isotropic APC-2/AS4 Composites	Stainless steel	Optimization using short-beam shear test is conducted to identify the better welding parameter Optimal parameters are found for the welded specimens
Lu et al. (1991) [85]	Eight-harness stain weave glass fabric reinforced PEI composites	Stainless steel (AISI 304L) metal mesh and M 200 metal mesh	Analysis of strength and failure mode is carried out. In failure mode, interlaminar failure is the major defect which is obtained from the results
Truckenmüller et al. (2006) [37]	GF/PP	CF stainless steel	Taguchi method is used to analyze the optimum process parameters. ANOVA is aided to analyze the individual parametric effect on the output

Table 2.5 Prominently encountered issues and remedies

Welding method	Prominently encountered issues	Remedy
Resistance welding	Spatter or molten material being expelled out during welding operation	Appropriate tip alignment will minimize the spatter. Decreasing the tong pressure and excessive pressure reduces the expulsion of molten material
	Poor weld or no weld at tips	Reduction of the tong length will provide strong weld at tips. Strong weld is accomplished by using standard thickness of the material for the machine
	Inconsistent weld nugget	Adopting the weld timer in the set up will yield consistent weld nugget

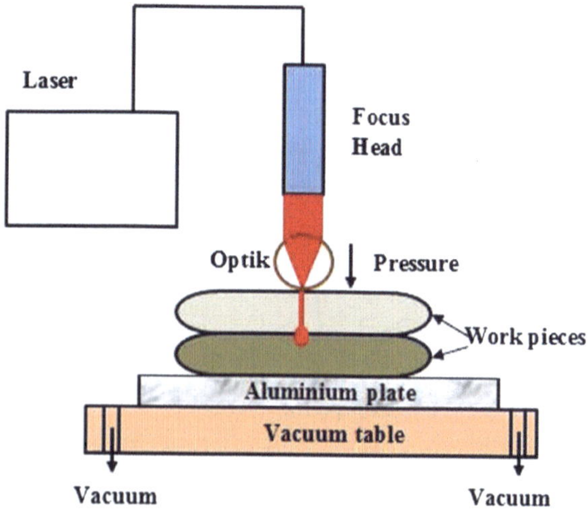

Fig. 2.11 Laser transmission welding technique [45]

Lasers used for materials pose complexity during joining compared to other techniques and in specific extended for hard to access regions and for exceptionally small parts. Inert gas shielding is essential for further reactive materials. Researchers presented test samples for inner pressure loading at the laser bond interface. Many experiments of S/N curves were conducted for inner pressure loading, and various laser weld configurations were established [45].

Laser welding of polymers is gradually increasing in industries like automotive, medical technology, textiles, consumer electronics, or packaging. Besides, it is an alternative for the polymer welding processes such as hot plate, ultrasonic, or vibration welding. Requirements on the quality of the joint, the elasticity of the production system, and the processing speed have raised [46].

Light scattering of the upper polymer comprises high persuasion on welding quality. Experimental and numerical computation techniques are evaluated for low-density polyethylene (LDPE) and light scattering of high-density polyethylene (HDPE). Initially, the beam quality of semiconductor laser is assessed; power flux allocation of the laser beam in a focused plane is examined by knife-edge technique. Subsequently, the power flux allocations of the laser beam subsequent to its pass-through LDPE/HDPE are evaluated by line scanning technique. Finally, collective mathematical model is used to compute scattering coefficient and standard variation of scattering, scattering-related factors, and the laser power flux allocation at the bond interface achieved [47].

The Bouguer–Lambert law (Eq. 2.9) is used to depict the laser power concentration in single-scattering or non-scattering polymers [48]

$$T = \frac{P_{out}}{P_L} = (1 - R_1)e^{-A_1 D} \tag{2.9}$$

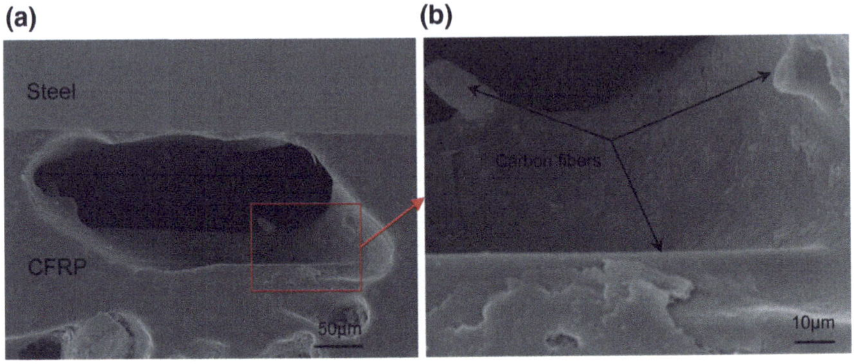

Fig. 2.12 SEM images of porosity I [16]

where P_{out} is the total laser power output after passing through the polymer specimen of thickness D, T is the fractional transmission, P_L is the incident laser power, and R_1 represents the factor that considers thickness-independent losses owing to reflection from the lower and upper surfaces in addition to backscattering from bulk material under the incident surface. The alterations generated by the defocused laser beam into the bulk material have been initially studied based on the laser process factors aspiring to generate incessant melting. Results have been assessed depending on the heat accumulation models. Lastly, laser welding of PMMA specimens has been productively revealed and evaluated by leakage tests for usage in direct laser assembly of micro-fluidic equipments [49].

2.4.8 Critical Process Parameters and Phenomena

Porosity I is the stage of the pore. Concurrent to CFRP pyrolysis kinetics under various bonding factors, while the heat input is greater than 77.8 J/mm, CFRP pyrolysis occurs near the central part of welding region during laser joining processing. The cross-link reaction among amide groups occurs initially resulting in dehydration at the stage of CFRP pyrolysis. Subsequently, hydrolysis of the amide connection happens, generating a huge number of gas products like NH_3, CO_2, H_2O, and hydrocarbons which guides to the formation of porosity I (Fig. 2.12). Hence, porosity I can be entirely suppressed by restricting the heat input to eliminate pyrolysis.

Porosity II has irregular geometries and rough surfaces (Fig. 2.13), which reveals shrinkage porosities produced under thermal shrinkage stress. Shrinkage porosities are generally produced as a result of solidification shrinkage in the CFRP manufacturing process [14, 15, 50–55].

Porosity II is the shrinkage porosity stage whose configuration is susceptive to the solidification succession; the backside cooling can create porosity II shift from periphery of melted region toward the interface. CFRP nearer to joint region solid-

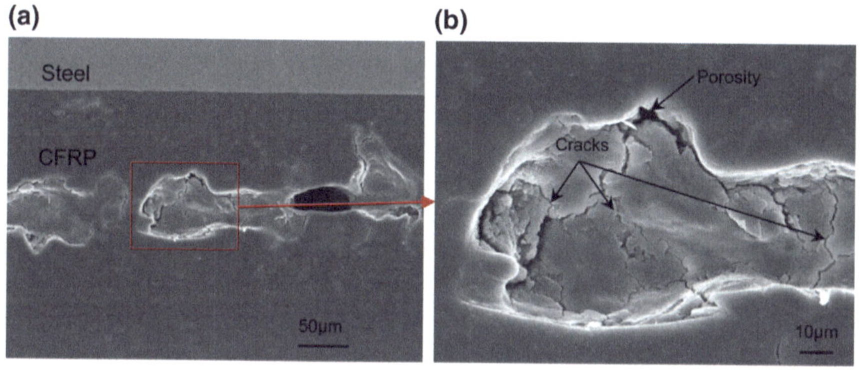

Fig. 2.13 SEM images of porosity II [16]

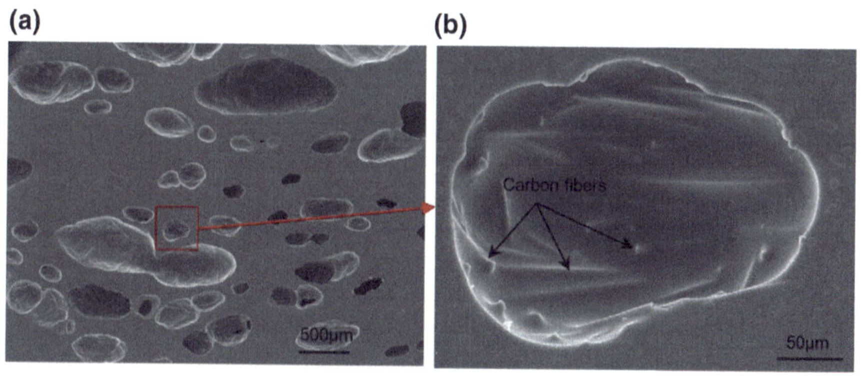

Fig. 2.14 Porosities in CFRP after constant temperature heating [16]

ifies at first due to the high thermal conductivity of steel at the stage of the laser bonding method, while CFRP far from joint region solidifies at last. The solidification shrinkage of CFRP leads to the configuration of porosity II, particularly in the last solidification region (Fig. 2.14).

2.4.9 Prominent Observations in the Laser Welding

Laser welding of multi-material classes has a wider potential in a variety of applications due to the non-contact nature of the process. This investigation evaluates the impacts of CO_2 laser process factors on the behaviors of spot bonds between thermoplastics and ceramics. The study reveals an increase in bond tensile strength with the rise in the number of spots and the time of exposure. Nevertheless, at low laser power, the impact of standoff distance on the spot weld diameter is obtained to be uncertain. Optimal process factors are examined using the multi-objective optimization on the origins of ratio analysis. The analysis reveals that the maximum number and maximum duration of laser spot light majorly alters the optimal behaviour of the joints [56].

Laser welding of transparent polymers is relatively an unexplored method, particularly with QSLW. In this investigation, it was revealed that quasi-simultaneous laser welding of transparent thermoplastic elastomer (TPE-A) was feasible. The impact of line distance and scanning direction was unimportant for the examined range. The maximum weld strength of 18.4 N/mm and the precise weld width were obtained with line distance of 0.2 mm. Greater strength and a wider weld were achieved with offsetting method as compared with conventional QSLW. For conventional QSLW, the weld width, height, and strength are obtained with 0.57 mm, 1.45 mm, and 14.4 N/mm, correspondingly. The weld width, height, and strength for offsetting technique were obtained with 0.77 mm, 0.99 mm, and 18.4 N/mm, correspondingly. The shape of the weld was the key significant variation among conventional QSLW and beam offsetting methods. Weld of the conventional QSLW is narrower when compared with beam offsetting method, and round shape of the weld is formed with offsetting method. For thin specimen welding, key parameter is the height of the weld. A little lower height of the welding for the offsetting method is obtained when compared to conventional QSLW method. Thus, beam offsetting method is utilized to weld thin material with no melting on the entire thickness of the upper plate. Beam offsetting process requires to be enhanced to obtain various types of heat distribution. More heat is required to weld at the sides of the specimen than the middle of the specimen. Aforementioned welding of the side of specimen is carried out by maintaining uneven line distances at the specimen [57].

Establishing mechanical interconnecting of polymer to metal at the weld interface was carried out by the metal pre-treatment using nanosecond pulsed laser that yields an accurate and reliable result. The enhancement of the mechanical interconnection which is the distance between adjacent grooves is achieved by the microstructure factor. A less important persuasion on the T-joint mechanical performance was obtained by the number of iterations (N tracks). The considered choice of joint region and the structuring direction showed negligible influence on the bond interface strength values. Initially, a dependence of internal cavity angle with number of iterations was verified. The internal cavity angles (45°) were attempted at higher number of iterations. Secondly, various failure modes were obtained while the internal cavity angle was close to 45°. However, additional bond strength enhancement was not recorded. The topographical and morphological aspects of the unfastened surfaces subsequent to pullout tests displayed that the volume of the various groove geometries (MC1-MC16) was entirely filled at the stage of the laser conductive joining process [58].

The laser concentration profile needs to be close to the optimal M-shape beam for improved weld seam quality that is optimized by the diffractive optical elements (DOEs). The results are provided for the enhanced weld seam width steadiness and broadening of process windows evaluated to flattop or Gaussian beam profiles. Arrangements in which the laser beam geometry can be utilized and optimized with no modification or realignment of the laser, optical head or fiber-optic cable are illustrated [59]. Literatures reporting various process parametric influences, types of materials being used and the corresponding results recorded during laser welding of polymers are listed in Table 2.6.

Initially, many metal pre-structuring techniques, containing selective laser melting (SLM), are demonstrated and their ability to offer undercut structures in the metal was evaluated. Secondly, the process factor ranges for hybrid joining of several metals and polymers are provided. Both direct and transmission laser welding processes are shown in this study. Clamping devices and optical heads particularly adopted for hybrid joining method are established. Broad lap-shear test results are illustrated that describe the bond strengths surpassing the base material strength generally attained with polymer–metal bonding [60].

Micromorphological observations of the weld region provided the reasons for the high welding strength in the laser joining of plastics which are typically the modifications of the polypropylene and chemical reactions of the plastics such as the TGMPP and PA66 proved for the higher strength. The joining efficiency of the solvent-based technique and the laser welding was evaluated and compared. Additionally, an impact modifier was used with the purpose to boost toughness of PLA and to enhance material processability.

Single-lap plastic bonds may be attained using laser welding by overlaying two plastic sheets with one laser transparent and other being absorbent. To enrich the absorption of polyethylene, crystalline carbon nanomaterials (filler materials) are used in this investigation. Nd:YAG laser working at 1064 nm was used as a source to form the joints. Strength of the bonds was evaluated using mechanical, morphological, and calorimetrical tests. To obtain the best weldment, the key focus is laid on the exposure time to laser light and filler amount. It is observed that in the optimal circumstances (with 0.2 wt% filler and 60 s exposure time) the joint depicts fair shear strength with a regular joint region [61].

Mechanical strength of polymer joint interfaces is studied by using transmission laser welding (TLW) for the high-density polyethylene (HDPE) and polypropylene (PP). Lap joint of the weld is formed with absorber (0.4 wt% of carbon black) which is evaluated for mechanical properties using instron tensile testing machine. From the earlier studies, the tensile strength (89%) of HDPE/HDPE welds was achieved in a tensile strength of PP/HDPE weld with HDPE used as the absorber. Greater weld strength is achieved for low reiteration time typically set at few milliseconds for PP and HDPE. Besides, it is also important to choose polymers with lower mesh size that possesses high crystallization temperature [62]. Majorly encountered problems in the laser welding and its solutions are given in Table 2.7.

2.4.10 Friction Stir Welding

Friction stir welding (FSW) is a solid-state process used for joining materials with various thicknesses, and it is the only choice for joining metal matrix composite [17]. A non-exhaustible rotating tool with a customized pin and shoulder is injected into

Table 2.6 Observations from laser welds

Author(s)	Methods	Polymers	Measured parameters	Results
Wang et al. (2009) [66]	Spectroscopy	Polypropylene (PP), polycarbonate (PC), polyamide 6 (PA6), and glass fiber-reinforced polyamide 6	Transmission	Specimen thickness revealed strong impact on scattering and absorbance. Scattering was also seen to rise with raising glass fiber content in PA6
Wang et al. (2010) [83]	Spectroscopy	Polycarbonate (PC) with different additives	Transmission Reflection	To attain scattering and absorption coefficient, a four-flux model of radiation transport in the diffusive estimation was utilized. Scattering coefficient is affected by adding additives
Aden et al. (2010) [67]	Spectroscopy	Polypropylene (PP), polyamide 6 (PA6), and polyoxymethylene (POM)	Transmission Reflection	Bouguer–Lambert law is utilized to compute absorption coefficient based on the material temperature. There is no important variation of absorption coefficient with temperature observed while comparing with amorphous thermoplastics and semicrystalline thermoplastics that demonstrated a unique reduction of absorption coefficient in the melting range as a result of the semicrystalline structure
Geiger et al. (2009) [84]	Power meter and spectroscopy	Polyethylene (PE) and polypropylene (PP) films (between 10 and 800 mm)	Transmission Reflection	The observation was made for a linear relation among natural log of transmission and sample thickness. Attenuation coefficient was based on the slope of this curve. The reflectance was roughly stable for the thickness among 10 and 800 mm

(continued)

Table 2.6 (continued)

Author(s)	Methods	Polymers	Measured parameters	Results
Coelho et al. (2004) [68]	Power meter and spectroscopy	Nylon 6 materials with short glass fiber or other fillers	Transmission	The laser energy transfer reduced monotonically with rising fiberglass content for nylon 6 plastics. The technique of spectroscopy proffered a lower transfer due to a huge fraction of the transferred light reaching the detector as compared to power detector
Kagan et al. (2002) [69]	Power meter	Polycarbonate (PC) and high-density polyethylene (HDPE)	Transmission Reflection	Reflectance was not based on thickness but increased with increasing laser incidence angle for both materials
Rhew et al. (2003) [70]	Power meter	Polyetheretherketone (PEEK) and polycarbonate (PC)	Transmission	Bouguer–Lambert law is used to calculate the attenuation coefficient. A linear relation among sample thickness and natural log transmission was seen, and the attenuation coefficient is developed by the slope. PEEK has a much superior attenuation coefficient as a result of the semicrystalline structure than amorphous PC
Dosser et al. (2004) [71]	Power meter	Polyamide 6 (PA6), glass fiber-reinforced polyamide 6 (PA6GF), and polycarbonate (PC)	Transmission	A linear relationship between the part thickness and the estimated natural log of the polymer transmittance for the light scattering and non-scattering polymers PA6, PA6GF, and PC was observed
Chen et al. (2011) [72]	Power meter	Polyamide 6 (PA6) and polycarbonate (PC) reinforced with 10% glass fiber	Transmission	The transmission lessens as the laser angle of incidence increases for a provided thickness. The transmission reduces as the thickness increases for various laser incidence angles

(continued)

Table 2.6 (continued)

Author(s)	Methods	Polymers	Measured parameters	Results
Azhikannickal et al. (2012) [73]	Thermal imaging technique	Polyamide 6 (PA6)	Reflection distribution of reflected laser light	Thermal imaging method displayed concurrences between the assessed total power and the actual laser input power, and among assessed power allocation, and it is examined experimentally through the knife-edge-based beam profiling method
Azhikannickal et al. (2012) [74]	Power meter	Polycarbonate (PC), short glass fiber-reinforced PC, polyamide 6 (PA6), polyamide 66 (PA66), and polypropylene (PP)	Transmission	The linear relationship among the estimated log of transmission and specimen thickness was observed. The result was utilized to attain an apparent absorption coefficient which detains thickness-dependent transmission losses and a noticeable reflection which was recorded for thickness-independent losses
				Roughly, linear relationship among clear absorption coefficient and glass fiber volume fraction was recorded
				The clear reflection firstly reduced and then increased with the glass fiber concentration for all glass fiber-reinforced materials. The clear absorption coefficient and clear reflection both were recorded that rises with the crystallinity as a result of the raised scattering

(continued)

Table 2.6 (continued)

Author(s)	Methods	Polymers	Measured parameters	Results
Xu et al. (2015) [75]	Metallographic	AISI 304, HexFlow RTM6	Transmission, oscillation width, oscillation frequency, and oscillation mode	It is significant to prevent a simultaneous rise of the seam width due to the rise in the amount of evaporated contagion material. From the observation, the maximum reduction of porosity was recorded for oscillation of frequency of 400 Hz. Seam porosity is identified from longitudinal sections, and it has been lessened from more than 50 to 20% by adopting beam oscillation
Woizeschke et al. (2015) [86]	Thermal imaging technique	CFRP panels with a polyamide resin (PA6/T) and mild steel sheet	Transmission, laser power, and scanning speed	Thermal degradation of CFRP is affected by the heating rate. Porosity I is allocated nearer to the interface, specially focused at the central area of melted region, with a smooth inner wall and regular shape. These porosities initially become visible, while the heat input is greater than 77.8 J/mm. Porosity II with roughening of the inner wall and an irregular shape is allocated a specific distance from the welding region, nearer to the periphery of the melted region
Tan et al. (2015) [88]	Metallographic and optical imaging	PMMA and titanium	Transmission HAZ and melt zone	The good joint is produced with laser power of 25 W, beam intensity of 70 W/cm^2, and processing time of 15 s. As estimated, both the melted region and the heat-affected region raise with the laser power. The results for processing time beyond 25 s were not achieved due to substrate deformation for all laser power levels

(continued)

Table 2.6 (continued)

Author(s)	Methods	Polymers	Measured parameters	Results
Jiang et al. (2015) [89]	Optical imaging	Polyamide 6 (PA6), Grafil TRH50 60 k	Transmission, HAZ	One of the major significant parameters persuading the bonding method is the applied material and its optical and thermal properties. Significant variation among carbon-reinforced and carbon-unreinforced polymers is the changing thermal conductivity. This is attributed to the fact that the thermal conductivity of carbon fibers is inhomogeneous, anisotropic, and much greater than that of the boundary polymer matrix that uses fast velocities yielding considerable drop of the seam strength data. Another significant problem is the temperature allocation of the fiber orientation. The temperature can be allocated and enhanced along the fibers into the polymer volume at the upper welding specimen. Enhancement of the heat allocation and thus the obtainable weld strength may be accomplished by lessening the necessitated energy per unit with the change in optical properties of the matrix polymer by including carbon block as absorbing elements
Cosson et al. (2015) [90]	IRFT spectrometer, inverse method	PC with glass fiber composite	Transmission factor and reflection factor	The specific heat capacity of the material gets modified due to the increase in the glass transition temperature (Tg) of the amorphous polymer (PC) with glass fiber composite which is 135 °C. The light scattering coefficient value attained is $D_\lambda = 0.940\ \mu m = 740\ m^{-1}$

(continued)

Table 2.6 (continued)

Author(s)	Methods	Polymers	Measured parameters	Results
Berger et al. (2015) [91]	Response surface methodology	PC-ABS	Weld strength, track with	Laser power plays the significant role creating impact on the weld strength, followed by the welding speed and standoff distance. Clamp pressure has insignificant impact on weld strength. Standoff distance is the key parameter creating impact on the track width, followed by the welding speed, laser power, and clamp pressure
Asséko et al. (2015) [92]	Response surface methodology, finite element method (FEM)	Polyethylene terephthalate (PET) and polypropylene (PP)	Molten pool depths, shear strength	The prediction of the weld width and molten depths is examined by the RSM. The precision of the models is validated with the interlinked p-values which are lower than 0.05. The molten depths and weld width rise in the entire design space with greater laser power, lesser welding speed, and lesser standoff distance. The current finite element model is competent of envisaging the molten pool geometry with a fair degree of accuracy and has higher importance in directing the LTW experiments with agreeable precision
Acherjee et al. (2015) [93]	Thermal and optical	ABS and CNT	Transmission and reflection	The reflection and absorption coefficients rely on the CNTs content. Therefore, the laser energy needed for generating a specific welding seam reduces when the CNTs content rises. For lesser CNT concentration, the process increases considerably and the overlapping zone among both regimes was very tiny. ABS with more CNT concentration was highly susceptible to differences in the input power

(continued)

Table 2.6 (continued)

Author(s)	Methods	Polymers	Measured parameters	Results
Wang et al. (2014) [94]	Thermal and optical	PC	Transmission	Rise in the data of weld width and weld strength with laser power was recorded. Likewise, receptivity of clamp pressure and standoff distance on weld width and weld strength is positive. Welding speed has unconstructive values of sensitivities for weld width and weld strength; rise in welding speed provides a reduction in the weld width and weld strength. Weld strength is controlled by the welding speed
Rodríguez-Vidal et al. (2014) [95]	Anodizing treatment	Polyethylene terephthalate (PET) and aluminum alloy (A5052)	Shear strength and joining behavior	Rising pulse duration and heat input is offered to increase the shear strength. Molten pool depth would persuade the shear strength as a result of the shear deformation. The shear strength for the welded sample with anodized bonding interface was considerably greater than that of the welded sample with obtained bond interface
Acherjee et al. (2014) [96]	Optical microscope	ABS	Transmission, shear strength	Optical microscope investigations offer weld seams with good exterior and weld widths among 0.8–1.5 mm and 0.8–1.1 mm for high and low concentration of CNTs correspondingly. ABS joining is based on the microstructure and optical properties of the material

(continued)

Table 2.6 (continued)

Author(s)	Methods	Polymers	Measured parameters	Results
Yusof et al. (2012) [97]	Artificial neural network (ANN)	Acrylic sheets	Shear strength and weld seam with	With ANN, the best architecture which is back-propagation algorithm with adaptive learning rate (4-4-5-2) is achieved. ANN model demonstrates a better prediction rate as compared to other MRA models
Rodríguez et al. (2012) [98]	Taguchi method with gray relational analysis	Acrylic sheets (PMMA)	Transmission, weld strength, and weld width	This optimization process switches the variety of quality behaviors of laser transfer joining method to a single-performance behavior named gray relational grade and therefore minimizes the optimization procedure. Laser power influences the output to the maximum difference and followed by welding speed and defocal position. Weld width and weld strength can be enhanced concurrently by the proposed technique
Acherjee et al. (2011) [99]	Spectroscopy	PEEK	Transmission	Lower laser speed (4 and 8 mm/s) offers the maximum weld strength. Amorphous PEEK was sensitive to heat damage at the superior power, while no heat damage was seen in semicrystalline PEEK. Amorphous PEEK weld strengths were lower than the semicrystalline PEEK weld strengths at all power settings and speeds. Bubbles existed in the weld interface in both amorphous and semicrystalline PEEK; this may be due to the vaporization of absorbed water molecules inside the PEEK

(continued)

Table 2.6 (continued)

Author(s)	Methods	Polymers	Measured parameters	Results
Acherjee et al. (2011) [100]	Infrared thermography	PMMA-ABS/PC	Transmission	A contactless technique like the IR thermography is used for surface temperature measurement on a distinctive arrangement for the laser welding of polymers. Internal flaws cause vital rise in temperatures which is detected through IR thermograms (small white region appears in the weld interface)
Amanat et al. (2010) [60]	Ray tracing	Glass and thermoplastic	Transmission, optical and thermal properties	Adequate heating of the polymer for melting takes place in a localized region at the materials' interface. The numerical technique computed follows the laser beams' optical path in incessant fiber composite and identifies the significant parameter for joining process optimization: heating time and joint face temperature field. The offered numerical technique is restricted to assembly of UD plies, but can be widened and in addition can improve various fabric structures, like woven fabrics. This observation is attributed to differences of fiber orientation in the thickness path persuaded by the weaving method

Table 2.7 Prominently encountered issues and remedies

Welding method	Prominently encountered issues	Remedy
Laser welding	Improper focal distance in contour welding	Setting the suitable distance between the focal point of the laser optic and the workpieces will yield better joints
	Poor weld strength in contour welding	High power density reduces the weld strength and influences the material to decompose. Low power density causes lack of fusion and weakens the weld. Hence, appropriate power density will provide better weld strength

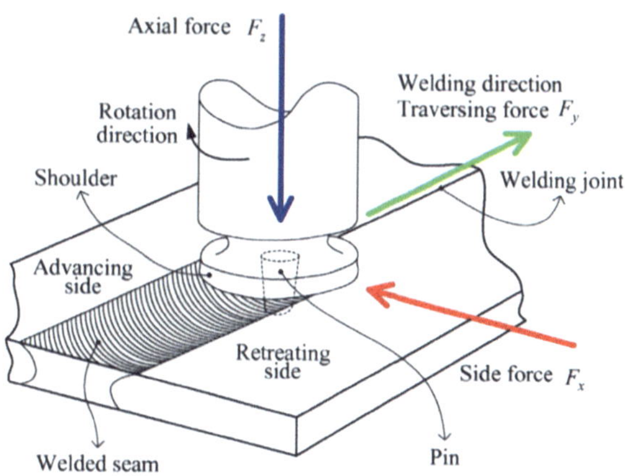

Fig. 2.15 Schematic representation of the FSW process [3]

the abutting edges of specimens to be welded and traversed along the line of the bond as displayed in Fig. 2.15. The tool generates the heat by traveling transversely into the specimen and creating the contact among the shoulder and the specimen. Tool travel into the specimen forces plastic deformation of the specimen in the stir region. The base metal experiences the high heat energy and strain at the time of stirring which causes dynamic recrystallization, making the finer grains in the joint region [27]. FSW process is illustrated in the schematic diagram (Fig. 2.15).

Material transfer occurs by directing the tool to the tracking edge, where it forms a bond between two workpieces. FSW technique may present several merits over current production methods, such as increased output quality as a result of ease of mechanization and feasibility of lengthy continual welds, with an ability to join all thermoplastics. Over the merits, the FSW method produces linear joints. FSW of polymer varies with that of metals due to variations in material structure and morphology. Transition region between the seam and the base material moves from the bottom of the tool pin in retreating face which are acceptable, wherein investigations

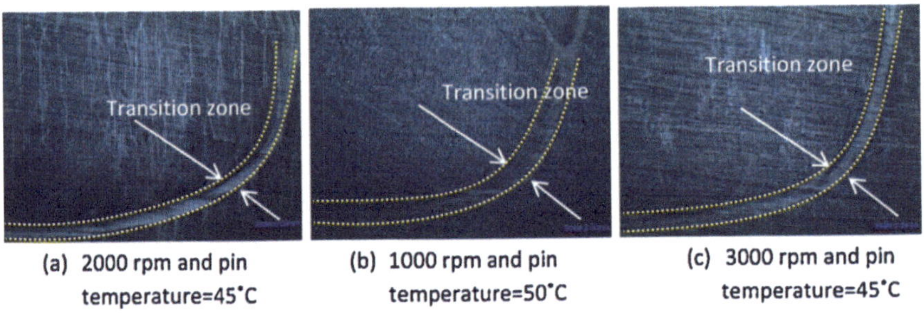

Fig. 2.16 Transition zone at different welding conditions [28]

revealed that the strength of the FSW weld was nearer to that of the base material. The study revealed good intensity in the morphology, while the width of the transition region is narrow (Fig. 2.16). Cooling, molecular relaxation/alignment, and crystallization are key parameters to limit the width of the transition region where the crystallization plays a vital role [28].

Mechanical properties are investigated for bonded specimens using standard tensile test as per ASTM D638. ANOVA of the regression models was studied to examine experimental outcome. The optimization method like response surface methodology is utilized to obtain the tensile strength properties of the welded joints. Conical pinned tool appears to be precise than the cylindrical pinned tool to envisage the weld strength from the two models which revealed from Table 2.8 [30]. Illustrations of the MX Triflute diagram in Table 2.8 reveal the difference of the obtained signals from torque (Mz), plunging force (Fz), temperature of the tool (TT), and temperature of the material (TM) measured at the bottom surface of the lower sheet [43]. With an increase in the tool, rotational speed more than 1500 rpm adversely impacts the weld quality other than creating alternatives in energy and thereby lessening the performance of the process [56].

2.4.11 Heat Generation

The total heat generated in the FSW (Eq. 2.10) can be computed using the equation ($Q_{total} = Q_1 + Q_2 + Q_3$). The values Q_1, Q_2, and Q_3 are heat generated under the tool shoulder, the tool probe side, and at the tool probe tip [36].

$$dQ = \omega dm = \omega r dF = \omega r \tau_{contact} dA \tag{2.10}$$

where M is the moment, A is the contact area, F is the force, and r is the cylindrical coordinate. The analytical model to forecast the heat generation excludes the strain rate that depends on yield shear stresses, non-uniform pressure distribution and the material flow governed by threads or flutes. Shoulder heat generation is attained by

Table 2.8 Analysis of tools used in FSW [29]

Tool	Tool pin shape	Ratio of pin volume to cylindrical pin volume	Swept volume-to-pin volume ratio	Rotary reversal	Application	Schematics
Cylindrical	Cylindrical with threads	1	1.1	No	Successful for butt welding; fails in lap welding	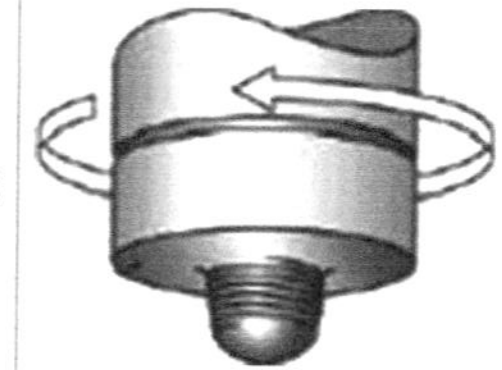
Whorl	Tapered with threading	0.4	1.8	No	Butt welding with lower welding torque	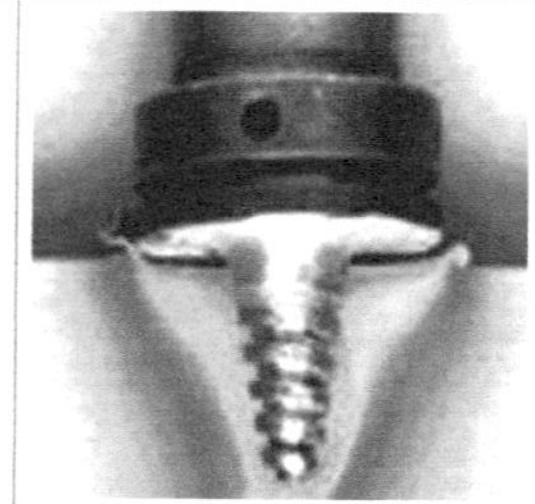

(continued)

Table 2.8 (continued)

Tool	Tool pin shape	Ratio of pin volume to cylindrical pin volume	Swept volume-to-pin volume ratio	Rotary reversal	Application	Schematics
MX Triflute	Threaded, tapered with three flutes	0.3	2.6	No	Butt welding with further lower welding torque	
Flared Triflute	Triflute with flute ends flared out	0.3	2.6	No	Lap welding with lower thinning of upper plate	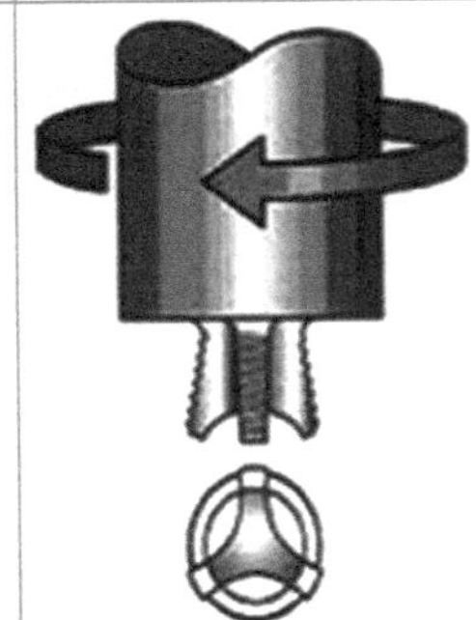

(continued)

Table 2.8 (continued)

Tool	Tool pin shape	Ratio of pin volume to cylindrical pin volume	Swept volume-to-pin volume ratio	Rotary reversal	Application	Schematics
Askew	Inclined cylindrical with threads	1	Depends on pin angle	No	Lap welding with lower thinning of upper plate	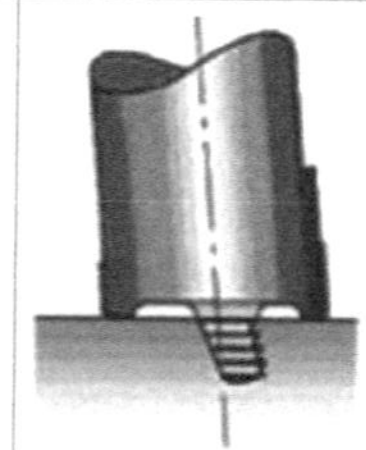
Re-stir	Tapered with threads	0.4	1.8	Yes	When minimum asymmetry in weld property is desired	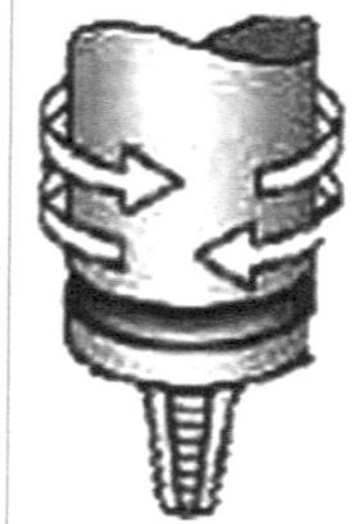

integrating the shoulder region from R_{probe} to $R_{shoulder}$ using the (Eq. 2.11) equation [63].

$$Q_1 = \int_0^{2\pi} \int_{R_{probe}}^{R_{shoulder}} \omega \tau_{contact} r^2 \,(1 + \tan\alpha) dr d\theta$$
$$= \frac{3}{2}\pi\omega\tau_{contact}(R^3_{shoulder} - R^3_{probe})(1 + \tan\alpha) \quad (2.11)$$

The distribution of the heat generated at the pin surfaces and the tool shoulder is assessed by the welding variables and the tool geometry. The local variations in the heat generation rates lead to differences of temperature on the tool surface. Heat transfer is considered three-dimensionally due to plastic flow fields and meaningful modeling of temperature [64]. The performance of friction stir welding with multi-material joining is compared with other welding methods. Nevertheless, this method has the merits like no special pre-treatments required before welding of materials [65]. Rotational speed, geometry, traverse speed, and tilt angle of tool are the welding variables for analysis. Indentation creep tests were conducted on bond joints utilizing a constant load (1 kg f) with varying time (0.5, 2, 5, 15, 30, 60, 90, and 120 min). Also, classical power law, Tabor, Juhasz, Ashby–Sargent, and Mulhearn were used for computing stress proponent data as a criterion for creep assessment (Table 2.9).

Weld strength was attained in CF-PA66 FSW joints with proper machined surface; the deficiency of volumetric defects and lap-shear strength was recorded up to 26.8 MPa [37]. The metals were built up with an impact during heating and subsequent stirring due to their good thermal conductivity. Nevertheless, the polymer components were built up based on the impact by the FSW during stirring and subsequent heating due to their poor thermal conductivity. In this FSW process, earlier to the preheating, the macromolecular structure was completely split [66].

Addition of the MWCNTs in the materials enhance its elongation and tensile strength but reduces the hardness of the weld [67]. The microstructure of FSW specimens exhibits the characteristics of the weld in Fig. 2.17. FSW method is utilized to weld a nylon 6 plate of 10-mm thickness, particularly modeled with left-hand threaded tool pin profile. Nylon 6 polymers were welded using FSW with welding feed at 10 mm/min and tool rotational speed at 1000 rpm. Schematic diagram (Fig. 2.18) illustrates the impact of the weld formation with the rotation of the threaded pin profile in clockwise and counterclockwise direction. The purpose of the investigation was to obtain the impact of the tool direction for lessening the weld flaws. The study revealed that the FSW-fabricated weld with counterclockwise direction of the tool provided flawless joints with good material properties.

Various categories of FSW tool pin profiles like taper, cylindrical, triangular, square, grooved, and spiral are broadly used based on the application. Researcher observed poorest strength with the straight cylindrical pin profile [68]. Nylon 6 plate welded with FSW in the clockwise tool rotation offered poor contacting of chunk with base materials, large number of cavity, and semisolid-state material deposition

[69]. Material flow on retreating and advancing sides moderately varied [70]. On the retreating side, the material loss occurs due to the increased brittleness in this region. The traverse speed does not affect the joint quality for most of the plastic materials like HDPE and PP.

2.4.12 Indentation Response

It is complex to evaluate Vickers test results due to materials' (HDPE) lower hardness. The optical inspection of indent sizes is used with instruments of load–displacement indentation evaluation. The strength does not match due to the softening of cold-worked material that occurs using grain growth subsequent to static and dynamic recrystallization that delivers heat at the time of welding [71]. The increase in the rotational speed and traverse speed reduces the tensile strength and subsequently increases the tensile strength typically achieved with minimization of the rotational and traverse speed. The reduction in the tensile strength played the key factor in producing the air bubble (Fig. 2.16) and crack which eventually causes the reduction in the crystalline content [72].

Table 2.9 Methods for calculating the stress exponent [85]

Method's name	Equations	Explanations
Classical power law	$d = At^{\frac{1}{2n+1}}$	d—average diagonal length A—constant n—stress exponent t—time
Juhasz	$\dot{d} = A_1\left(\frac{F}{d^2}\right)^n$	$\dot{d}$—rate of indentation length variation with time A1—constant F—applied load d—average diagonal length n—stress exponent
Ashby–Sargent	$H_v(t) = \frac{\sigma_0}{(nC_4\,\dot{\varepsilon}\,t)^{\frac{1}{\pi}}}$ The slope of the ln (H_v) versus ln(t) plot is −1/n	Hv(t)—hardness as a function of the time σ_0—reference stress at which the strain rate is ε_0 C_4—constant (function of indenter shape geometric constants and n) t—time n—stress exponent
Mulhearn–Tabor	$-(n+0.5)\log H_v = \log t$ The slope of the log(Hv) versus log(t) plot is 2 / −2n + 1	n—stress exponent Hv(t)—hardness as a function of the time t—time

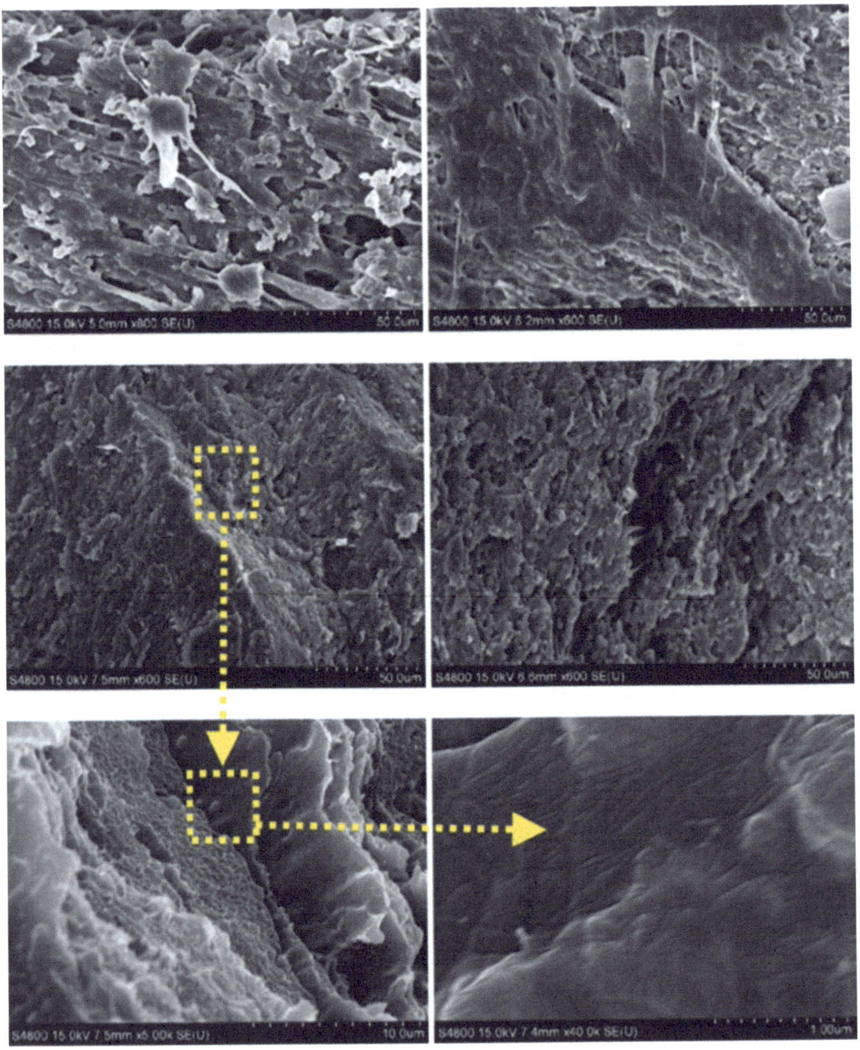

Fig. 2.17 SEM images of FSW specimens [83]

2.4.13 *Critical Process Parameters and Phenomena*

The joining of materials with increased rotational speed and reduced tool travel speed enhances the flexural strength of the joint and also lessens the flaws. The optimum welding factors to obtain maximum flexural strength was recorded at traverse speed of 25 mm/min, rotational speed of 1400 rpm, and shoulder temperature of 100 °C. In the aforementioned optimal circumstances, the flexural strength of welded specimen was about 66% of parent material strength [73].

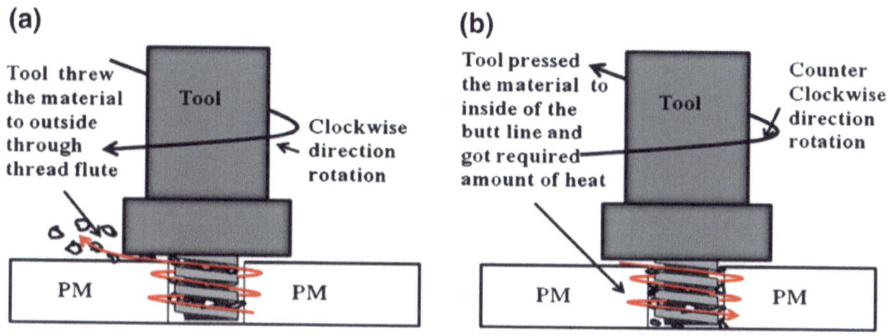

Fig. 2.18 Effect of the tool rotation in clockwise and counterclockwise direction [84]

The varying factors of process that involves the tool plunge rate, dwell time, and tool rotational speed significantly influence the weld strength. The significant factor was found to be the tool plunge rate. Major impact diagrams achieved by statistical analysis indicate that the weld strength reduces with increasing tool plunge rate. The higher and lower tensile strength of the welded samples was compared [74]. The weld quality and weld strength were enhanced with an increase in the rotational speed and the hot shoe temperature and in parallel reducing the tool traverse speed [75].

In the friction stir spot welding (FSSW), the general idea is the use of an accelerating agent at the welding interface of the polymer, thereby trying to create good bonds swiftly. The significant solution studied initiated the use of a plasticizer, which is generally used to lessen the viscosity of PEEK and enhance the chains' mobility.

2.4.14 *Prominent Observations in the Friction Stir Welding*

Researcher observed poorest strength with the straight cylindrical pin profile [76]. Nylon 6 plates welded with FSW in the clockwise tool rotation offered poor contacting of chunk with base materials, large number of cavity, and semisolid-state material deposition [77]. Material flow on retreating and advancing sides moderately varied [78]. On the retreating side, the material loss can take place which raised the brittleness in this region. The traverse speed does not affect the joint quality for most of the plastic materials like HDPE and PP.

It is complex to evaluate Vickers test results due to materials' (HDPE) lower hardness. The optical inspection of indent sizes is utilized with instruments of load-displacement indentation evaluation. The strength does not match due to the softening of cold-worked material that occurs using grain growth subsequent to static and dynamic recrystallization that delivers heat at the time of welding [78]. The increase in the rotational speed and traverse speed reduces the tensile strength and subsequently increases the tensile strength typically achieved with minimization of

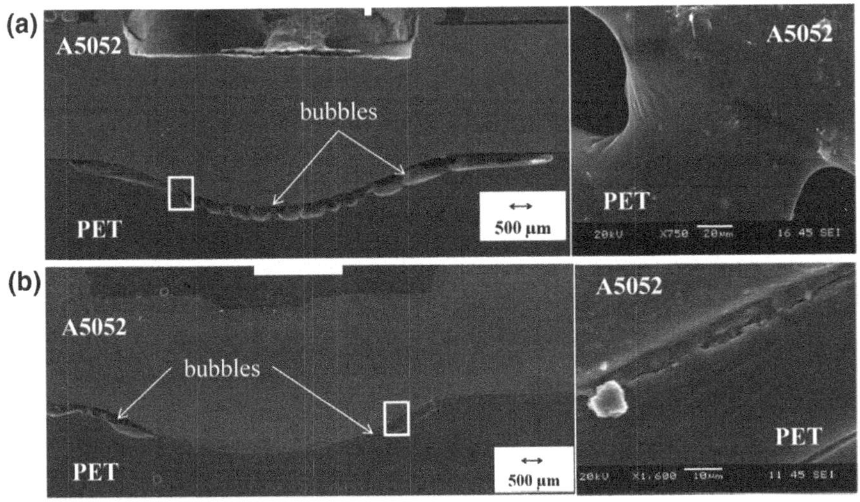

Fig. 2.19 Cross-sectional observation of A5052/PET joint [79]

the rotational and traverse speed. The reduction in the tensile strength played the key factor in producing the air bubble (Fig. 2.19) and crack which eventually causes the reduction in the crystalline content [79]. Friction stir spot welding (FSSW) has limitations; viz., spot geometries and overlap configurations can only be formed, FSSW is not suitable for thermoset matrix composites, and it is not suitable for very thick metallic parts (Table 2.10).

The joining of materials with increased rotational speed and reduced tool travel speed enhances the flexural strength of the joint and also lessens the flaws. The optimum welding factors to obtain maximum flexural strength were recorded at traverse speed of 25 mm/min, rotational speed of 1400 rpm, and shoulder temperature of 100 °C. In the aforementioned optimal circumstances, the flexural strength of welded specimen was about 66% of parent material strength [57].

The varying factors of process that involves the tool plunge rate, dwell time, and tool rotational speed significantly influence the weld strength. The significant factor was found to be the tool plunge rate. Major impact diagrams achieved by statistical analysis depict that weld strength reduces with increased tool plunge rate. The higher and lower tensile strength of the welded samples was compared [80]. The weld quality and weld strength were enhanced with increase in the rotational speed and the shoe temperature and in parallel reducing the tool traverse speed [81].

In the friction stir spot welding (FSSW), the general idea is the use of an accelerating agent at the welding interface of the polymer, thereby trying to create good joints rapidly. The significant solution studied initiated the use of a plasticizer, which is generally utilized to lessen the viscosity of PEEK and enhance the chains' mobility. Vacogne et al. (2011) investigated the modifications in the joint region microstructure using differential scanning calorimetry and scanning electron microscopy [82]. The

Table 2.10 Significant conclusions on FSW of polymer from a various literature

Author(s)	Materials	Rotational speed (rpm) and tool traverse speed (mm/min)	Parameters	Method of analysis	Remark
Khodabakhshi et al. (2014) [101]	Acrylonitrile butadiene styrene (ABS) sheets	800/20 1250/40 1600/80	Rotational speed, traveling speed, and shoe temperature at the beginning	Analysis of variance (ANOVA), response surface method (RSM)	Weld quality and weld tensile strength were enhanced with reduced tool travel speed and increased rotational speed
					Brittle failure occurs due to lack of material on the retreating side of the joint
					The tensile strength is based on the processing parameters for the polymer materials
Gao et al. (2014) [102]	Polycarbonate sheets (PC)	900 1500 2150	Dwell time, tool plunge rate, and rotational speed	Mechanical characterization and dimensional analysis of the joints	The mechanical behavior of the joints is positively influenced by reduced plunging speed
					Increasing the tool rotational speed is affected by reducing the processing forces, while the material mixing increases with higher temperature
					The dwell time creates impact on the material temperature, mechanical behavior of the joint, and dimension of the joint zone

(continued)

Table 2.10 (continued)

Author(s)	Materials	Rotational speed (rpm) and tool traverse speed (mm/min)	Parameters	Method of analysis	Remark
De Saracibar et al. (2014) [103]	Polypropylene (PP) composite	630/12 800/16 100/20 1250/25	Tool rotational speed, welding speed, tilt angle, and tool pin geometry on tensile shear strength	ANOVA, Taguchi method	Greater weld strength was obtained with threaded cylindrical–conical tool
					The major parameters governing the FSW are the welding speed, the rotational speed, tilt angle, and tool pin geometry
					The key-contributing parameter for the weld strength was rotational speed of 47.21%. The least contribution parameter for the weld strength was tilt angle of 5.65%
Azarsa et al. (2014) [104]	AA 7075 aluminum alloy and polycarbonate (PC) plates	3000 to 3500/50 to 150	Rotational speed, traveling speed	Microstructural analysis	The transfer of AA 7075 was extremely affected by welding factors and peak temperatures
					EDX and XRD analysis disclosed that no prominent mixing takes place between PC and AA 7075 and no ceramic-type compound was produced at the bond interface
					Hardness and tensile strength of the bonded joints reduced while compared with parent materials

(continued)

Table 2.10 (continued)

Author(s)	Materials	Rotational speed (rpm) and tool traverse speed (mm/min)	Parameters	Method of analysis	Remark
Ahmadi et al. (2014) [105]	Acrylonitrile butadiene styrene (ABS)	1000/50 1500/200	Rotational speed, traveling speed, and axial force	Influence of the parameter axial force	While maintaining the axial force and tool rotational speed over specific conditions, good quality joints were achieved
					Over those conditions, the production of porosities and cavities in the retreating side of the stir region is eliminated and the weld zone was even and smooth.
					The tensile strength and strain of joints were always lower than the parent material
Dialami et al. (2013) [107]	Polypropylene (PP) sheets	900/105	Tool pin geometries, pin angles, pin lengths, shoulder diameters, and shoulder angles	Lap-shear fracture load, weld stir zone formation	The threaded tool provided the highest tensile strength
					Weld lap shear fracture and stir region production were altered by the weld tool geometry
					The main significant factor for the weld strength and the weld quality are the pitch length of threaded pins

(continued)

Table 2.10 (continued)

Author(s)	Materials	Rotational speed (rpm) and tool traverse speed (mm/min)	Parameters	Method of analysis	Remark
Dashatan et al. (2013) [108]	Polypropylene (PP)	400/8 630/16 1000/20	Pin geometry, tool rotational speed, work linear speed, and tool tilt angle	Appearance and tensile strength were investigated	Tensile strength and weld appearance were affected by the tool pin geometry
					Tensile strength was reduced while increasing the work linear speed from 8 to 20 mm/min
					Tilt angle with 0 to 2° improved the tensile strength
Bagheri et al. (2013) [109]	High-density polyethylene (HDPE) sheets	710/25 1120/50 1400/100	Rotational speed of the pin, tool traverse speed, and shoe temperature	Response surface method (RSM), ANOVA	Enhanced weld flexural strength with little flaws achieved with reduced tool travel speed and increased rotational speed
					At optimal level, the flexural strength of welded specimen obtained was 66% of parent material strength
Bilici et al. (2012) [52, 115, 116, 125]	Acrylonitrile butadiene styrene (ABS)	900/6 1400/16 1800/25	Rotational speed, linear speed, tilt angle, shoulder diameter, pin diameter, diameters ratio, pin profile	Statistical methods, ANOVA, RSM	Enhanced welding strength is achieved with conical pinned tool when compared to cylindrical pinned tool
					Linear speed of welding with high speed has significant impact on joint strength

(continued)

Table 2.10 (continued)

Author(s)	Materials	Rotational speed (rpm) and tool traverse speed (mm/min)	Parameters	Method of analysis	Remark
Vacogne and Wise (2011) [55]	Carbon fiber-reinforced polyamide 66 laminate (CFPA66)	3000/-	Feasibility	Mechanical and optical microscopy	Joints for CF-PA66 materials were obtained with good surface finish, without volumetric flaws, and lap-shear strength up to 26.8 MPa
Saeedy et al. (2011) [120]	PMMA, ABS	500/8 800/16 1250/24	Tool rotational speed, tool plunge rate, and dwell time	Design of experiment (DOE)	Weld strength was influenced by the parameters such as tool plunge rate, dwell time, and tool rotational speed
					Tool plunge rate is observed as the key factor
					Increasing the tool plunge rate reduces weld strength, and rising the dwell time increases the weld strength
Bilici et al. (2012) [115]	Acrylonitrile butadiene styrene (ABS)	400/20 600/40 800/60	Shape of the pin, rotational speed, and translational speed	Mechanical property	Back slit and root defect of the welded specimens were avoided with newly designed tool
					Tensile strength is lessened with increased rotational speed
Payganeh et al. (2011) [122]	HDPE, ABS	2500/30	With and without the multi-walled carbon nanotubes	Microstructure analysis	Addition of the MWCNTs lessens the number of flaws such as crack and pores

(continued)

Table 2.10 (continued)

Author(s)	Materials	Rotational speed (rpm) and tool traverse speed (mm/min)	Parameters	Method of analysis	Remark
					Elongation and tensile strength were enhanced by including the MWCNTs, and it reduces the hardness of the joint
Azarsa et al. (2014) [104]	HDPE	1000 2000	Tool rotational speeds and tool pin temperature	Mechanical behavior and weld microstructure	The stir region reduces due to the ductile property of the bonds while increasing the tool pin temperature
					Turbulence of the material affected by the stirring action of the tool was the significant parameter that controlled the strength of the joints at reduced tool temperature
					FSW joint strength was nearer to the parent material when the width of the transition region is narrow lesser complex morphology
Gonçalves et al. (2015) [127]	Polyethylene (PE)	560–1120	Tool rotational speed, tool plunge depth, and dwell time	Mechanical property	The weld strength and chunk thickness were influenced by the tool pin geometry
					FSSW provided the good weld quality using tapered cylindrical pin FSSW influenced by the tool pin angle, the delay time, shoulder diameter, and the shoulder concavity angle

(continued)

Table 2.10 (continued)

Author(s)	Materials	Rotational speed (rpm) and tool traverse speed (mm/min)	Parameters	Method of analysis	Remark
Dashatan et al. (2013) [108]	Polyethylene (PE)	900/12.5 1120/20 1400/31.5	Rotational speed, traverse speed, geometry, and tilt angle	Creep evaluation	The enhanced stress elements were computed with various methods like Juhasz, Ashby–Sargent, classical power law, and Mulhearn–Tabor values
					Evaluation of the indentation creep characteristics was effectively carried out by the Mulhearn–Tabor method
					The increasing tool tilt angle produced lower weld quality and crack resistant of samples
Pirizadeh et al. (2014) [51]	Aluminum alloy AA6061 and MC nylon 6	2000/600	Tool rotation rate, welding speed, and dwell time	Optical microscopy	The rise of the rotation rate of the tool in the FSW increased the thermal input while bending the MC nylon 6 plates and AA6061 plates, reduction in the volume of bubbles, and an improvement in the NSS of the welds
					Raising the speed of the FSW lessened both the thermal input and the volume of bubbles
Gao et al. (2015) [47]	Nylon 6	1000/10	Tool rotation rate, welding speed	Threaded pin profile in clockwise direction and counterclockwise direction was analyzed	Joints with good properties were made from the left-hand threaded pin profile with counterclockwise direction of rotation

(continued)

Table 2.10 (continued)

Author(s)	Materials	Rotational speed (rpm) and tool traverse speed (mm/min)	Parameters	Method of analysis	Remark
Vijendra et al. (2015) [128]	Polycarbonate (PC)	1500/8 5400/46	Rotational speed, tool plunge rate, preheating time, dwell time, and waiting time	Artificial neural network (ANN), mechanical characterization	Processing times and processing speeds significantly influence the geometry and the mechanical characteristics of the FSSW joints
					The waiting time, the dwell time, and the tool plunge rate were the influencing factors for the weld strength of PC in FSSW. The key parameter is the dwell time which was observed
					The tool rotational speed had little impact on the weld strength, resulting in a negligible reduction of tensile shear strength
Bilici et al. (2012) [115]	Polypropylene (PP) sheets	2000/60 3000/60	Rotation speed, transition zone,	Optical and electron microscopy analysis	Slow cooling rate caused the spherulitic structure seen in the central part of the seam
					The heat-affected zone has showed with distorted spherulites under shear stresses where the PP became softer

(continued)

Table 2.10 (continued)

Author(s)	Materials	Rotational speed (rpm) and tool traverse speed (mm/min)	Parameters	Method of analysis	Remark
Hoseinlaghab et al. (2015) [129]	AA5059 alloy and high-density polyethylene (HDPE)	400 to 2000/ 30 to 200	Rotation speed, transition speed,	Mechanical characterization	Bonding mechanisms during welding method depicted mechanical interlocking through micro- and macro-constraints and interfacial chemical adhesion between the polymeric layers and aluminum
					Tensile fracture is conducted from combined region/AA5059 interface and denotes much greater tensile strength in the SZ than that of the AA5059/HDPE interface imputed to aluminum surface wettability
Liu et al. (2014) [130]	Acrylonitrile butadiene styrene (ABS)	1000 to 1500/50 to 200	Axial force, rotational and traverse speeds, and tool temperature	Mechanical characterization	Welds conducted by FSW on ABS plates utilizing a stationary shoulder tool and a robotic system
					Weld quality is enhanced with tool temperature of 115 °C and particularly the weld crown surface
					Higher axial force encourages the squeeze of the molten polymers, thereby avoiding introduction of air into the weld, and supports cooling of the weld, preventing shrinkage and voids' creation

(continued)

Table 2.10 (continued)

Author(s)	Materials	Rotational speed (rpm) and tool traverse speed (mm/min)	Parameters	Method of analysis	Remark
					Strain of welds and tensile strength were enhanced by the high axial force and high tool rotational speed
Panneerselvam and lenin (2014) [53]	Magnesium AZ31–O/glass fiber and carbon fiber-reinforced poly(phenylene sulfide)	900 3000	Rotational speed, tool plunge depth, joining time, joining pressure	Microscopic analysis	The presence of partial cracks is an indication of adhesive forces playing a vital role in bond formation
					Thermal aspects impact during bond consolidation, thereby influencing the material with cross-linking that reduces the degree of crystallinity
Lambiase et al. (2015) [131]	Polypropylene (PP) composite	800/16 1000/20 1250/25	Welding speed, rotational speed, and tilt angle	Taguchi method	Tensile shear strength was majorly influenced by the tilt angle and welding speed
					The rotational speed, optimum values of welding speed, and tilt angle were achieved for maximum tensile strength at 1250 rpm, 25 mm/min, and 1°, respectively
Kiss et al. (2012) [113]	HDPE, PA66	–	Temperature	Thermal analysis (DSC, TG)	The variation between these two plastic relays is due to the deviation of their macromolecular structures

(continued)

Table 2.10 (continued)

Author(s)	Materials	Rotational speed (rpm) and tool traverse speed (mm/min)	Parameters	Method of analysis	Remark
Khodabakhshi et al. (2014) [101]	HDPE	1500/45 2100/75 3000/115	Rotation speed, transition speed, tilt angle	Taguchi approach	The tool rotation speed acts as the key factor and contributes to the 73.85% of the overall welding factors. The tilt angle had minimal influence
					The welding factors were optimized for obtaining maximum tensile strength with tool traverse speed of 115 mm/min, the tilt angle of 3°, and the tool rotation speed of 3000 rpm
Mendes (2014) [132]	Aluminum alloy (A5052) to polyethylene terephthalate (PET)	3000	Spindle speed, plunge depth, plunge speed, holding time	Mechanical characterization, microscopic analysis	The plunge speed has a significant influence on the heat-affected zone
					The tensile shear strength for the joint sample with the plunge depth of 0.4 mm was noticeably lower than that for the joint sample with the plunge depth of 0.7 mm
					Increasing plunge speed lessens the tensile shear failure

Table 2.11 Prominently encountered issues and remedies

Welding method	Prominently encountered issues	Remedy
Friction stir welding	Loss of material in the joint line	Hot shoe will restrict the expulsion of the material and maintain the joint integrity. And also, it will be used as the rotating pin tool
	Non-homogeneous at the weld line	Hot shoe will be used to avoid expulsion and uneven mixing of plastics

observations of the microstructures were utilized to express the strength of the welded specimen. Major issues encountered in the friction stir welding and its solutions are given in Table 2.11.

2.5 Conclusions

Polymer joining has been carried out by various techniques such as hot gas welding, hot tool welding, resistance welding, laser welding, and friction stir welding. The critical observations reveal information on the process parameters, its strength, and quality issues. The influences of input parameters on the weld performance for various processes have extensively been reported. It may be emphasized that solid-state welding process is more appropriate as it accounts for the differential thermal gradient exhibited by polymeric materials expressed in terms of glass transition temperature. The weld characteristics are dictated by the transitions occurring in the weld regions that encompasses crystalline, amorphous, and semicrystalline natures. Mathematical tools have been used for modeling and simulation for forecasting thermal distributions, deformation pattern analysis, and parameter optimization. However, there is no concrete information available to justify certain claims such as

1. No information on effects experienced by polymer chain subjected to various joining processes;
2. Information on empirical relation existing between inter- and intra-dependent parameters has not been reported for polymer joining;
3. Inadequate reports on material flow behavior and the corresponding solidification;
4. Several tests may be conducted to evaluate the adhesiveness among the polymers for which standards are yet to be established;
5. No control and automation mechanisms have been designed or adapted for polymer joining process which may eventually minimize the error;
6. The usages of nondestructive test for exploring interfaces have not been presented for polymer joining;

7. Active and passive forces enforcing the diffusion of molecules have not been established with equations illustrating a prediction which may lack rationality;
8. The elastic phase and the transition associated remain unaddressed;
9. The static and dynamic characteristics occurring during the polymer joining process have remained undisclosed.

References

1. Dodin, M.G. 1981. Welding mechanisms of plastics: A review. *The Journal of Adhesion* 12 (2): 99–111.
2. Reitz, Wayne E., and Rebecca M. Oman. 2000. How to join plastics. *Advanced Materials and Processes* 158 (3): 49–52.
3. Marczis, B., and T. Czigany. 2006. Interrelationships between welding parameters of hot-gas welded polypropylene. *Polymer Engineering & Science* 46 (9): 1173–1181.
4. Balkan, Onur, Halil Demirer, Ayhan Ezdeşir, and Hüseyin Yıldırım. 2008. Effects of welding procedures on mechanical and morphological properties of hot gas butt welded PE, PP, and PVC sheets. *Polymer Engineering & Science* 48 (4): 732–746.
5. Schmachtenberg, Ernst, and Carsten Tüchert. 2003. Long-term properties of butt-welded poly (propylene). *Macromolecular Materials and Engineering* 288 (4): 291–300.
6. Ageorges, Christophe, and Lin Ye. 2001. Simulation of impulse resistance welding for thermoplastic matrix composites. *Applied Composite Materials* 8 (2): 133–147.
7. Amanat, Negin, Natalie L. James, and David R. McKenzie. 2010. Welding methods for joining thermoplastic polymers for the hermetic enclosure of medical devices. *Medical Engineering & Physics* 32 (7): 690–699.
8. Grewell, D., and A. Benatar. 2007. Welding of plastics: Fundamentals and new developments. *International Polymer Processing* 22 (1): 43–60.
9. Grewell, D., et al. 2003. *Plastic and composite welding handbook*. Munich: Hanser Publishers.
10. Tadmor, Z., and C. Cogos. 2006. *Principles of polymer processing*, 2nd ed. Wiley Inc.
11. Ratner, B.D. 2004. *Biomaterials science: An introduction to materials in medicine,* 2nd ed. Elsevier Inc.
12. Stokes, Vijay K. 1989. Joining methods for plastics and plastic composites: An overview. *Polymer Engineering & Science* 29 (19): 1310–1324.
13. Bahadir, Senem Kursun, Fatma Kalaoğlu, and Simona Jevšnik. 2015. The use of hot air welding technologies for manufacturing e-textile trasmission lines. *Fibers and Polymers* 16 (6): 1384–1394.
14. Diedrich, G., and B. Kempe. 1980. Welding of pipes and fittings made from different grades of HDPE. *Kunststoffe-German Plastics* 70 (2): 87–89.
15. Abram, J., D.W. Clegg, and D.V. Quayle. 1982. The strength of welds in uPVC. *Plastics and Rubber International* 7 (2): 58–60.
16. Hausdörfer, D., H. Herrmann, and W. Muth. 1976. Multi-layer hot-gas welding of thick-walled rigid polyethylene mouldings. *German Plastics* 66 (1): 4–6.
17. Atkinson, J.R., and B.E. Turner. 1989. Repairability of plastic automobile bumpers by hot gas welding. *Polymer Engineering Science* 29 (19): 1368–1375.
18. Stokes, Vijay K. 2000. A phenomenological study of the hot-tool welding of thermoplastics Part 2. Unfilled and glass-filled poly (butylene terephthalate). *Polymer* 41 (11): 4317–4343.
19. Stokes, Vijay K. 2001. A phenomenological study of the hot-tool welding of thermoplastics Part 3. Polyetherimide. *Polymer* 42 (2): 775–792.
20. Kim, H.C., D.G. Ahn, S.H. Lee, and D.Y. Yang. 2005. A study on thermal characteristics of non-contact hot-tool for rapid feature detailing (RFD) process. *International Journal of Machine Tools and Manufacture* 45 (3): 345–353.

21. Hou, M., L. Ye, and Y.W. Mai. 1999. An experimental study of resistance welding of carbon fiber fabric reinforced polyetherimide (CF Fabric/PEI) composite material. *Applied Composite Materials* 6: 35–49.
22. Harras, Bilal, Kenneth Chesley Cole, and Toan Vu-Khanh. 1996. Optimization of the ultrasonic welding of PEEK-carbon composites. *Journal of Reinforced Plastics and Composites* 15 (2): 174–182.
23. Bucknall, C.B., I.C. Drinkwater, and G.R. Smith. 1980. Hot plate welding of plastics: Factors affecting weld strength. *Polymer Engineering & Science* 20 (6): 432–440.
24. Taşkıran, E., S. Sayer, Ç. Özes, Ç. Yeni, and A. Ülker. 2015. Effect of process parameters and talc ratio on hot plate welding of polypropylene. *Materialwissenschaft und Werkstofftechnik* 46 (8): 860–872.
25. Liu, S.J., and H.F. Cheng. 2010. The influence of interface geometry on the joint strengths of hot plate welded composites. *Journal of Reinforced Plastics and Composites* 29 (4): 497–509.
26. Ülker, A., N. Öztoprak, S. Sayer, and C. Yeni 2017. Optimization of welding parameters of hot plate welded PC/ABS blends by using the Taguchi experimental design method. *Journal of Elastomers & Plastics* 0095244317740735.
27. Hessel, J., and E. Mauer. 1985. The appearance of fractures in welded joints made of plastics following bending tests. *Welding and Cutting* 37 (5): 76–77.
28. John, P., J. Hessel, and E. Gaube. 1985. A new development in the extrusion welding sector. *German Plastics* 75 (1): 7–8.
29. Bonten, C., and E. Schmachtenberg. 2001. A new hypothesis to describe the mechanisms acting in a welded joint of semicrystalline thermoplastics. *Polymer Engineering & Science* 41 (3): 475–483.
30. Sims, J., P.A. Ellwood, and H.J. Taylor. 1993. Pollutants from laser cutting and hot gas welding of plastics. *Annals of Occupational Hygiene* 37 (6): 665–672.
31. Marczis, B., and T. Czigany. 2006. Investigation of the heat affected zone of hot-gas welded PP joints. *International Polymer Processing* 21 (2): 141–148.
32. Stokes, Vijay K. 1998. Experiments on the hot-tool welding of three dissimilar thermoplastics. *Polymer* 39 (12): 2469–2477.
33. Stokes, Vijay K. 1999. A phenomenological study of the hot-tool welding of thermoplastics. Part 1: Polycarbonate. *Polymer* 40 (23): 6235–6263.
34. Xiao, X.R., S.V. Hoa, K.N. Street. 1994. Repair of thermoplastic resin composites by fusion bonding. In *Composites bonding*, vol. 1227, 30–44. ASTM STP.
35. Panneerselvam, K., S. Aravindan, and A. Noorul Haq. 2012. Study on resistance welding of glass fiber reinforced thermoplastic composites. *Materials & Design* 41: 453–459.
36. Shi, H., I. Fernandez Villegas, and H.E.N. Bersee. 2013. Strength and failure modes in resistance welded thermoplastic composite joints: Effect of fibre–matrix adhesion and fibre orientation. *Composites Part A: Applied Science and Manufacturing* 55: 1–10.
37. Truckenmüller, Roman, Ralf Ahrens, Yue Cheng, Günther Fischer, and Volker Saile. 2006. An ultrasonic welding based process for building up a new class of inert fluidic microsensors and-actuators from polymers. *Sensors and Actuators, A: Physical* 132 (1): 385–392.
38. Sims, J., P.A. Ellwood, and H.J. Taylor. 1993. Pollutants from laser cutting and hot gas welding of plastics. *The Annals of Occupational Hygiene* 37 (6): 665–672.
39. Atkinson, J.R., and B.E. Turner. 1989. Repairability of plastic automobile bumpers by hot gas welding. *Polymer Engineering & Science* 29 (19): 1368–1375.
40. Haque, M.S., M. Kumar, and A. Khan. 2015. Hot gas welding of plastic: Fundamentals & importance. *Indian Journal of Science and Technology* 8(4).
41. Greenawalt, E.L. 1985. Form, fill and seal machine with hot gas and thermal impulse sealing. Dow Chemical Co., U.S. Patent 4,512,138.
42. Toss, R., Toss Ramon. 2000. Process and device for hot gas welding of plastic sheets with electrically heated tubular gas flow conduit. U.S. Patent 6,134,387.
43. Hou, Meng, Mingbo Yang, Andrew Beehag, Yiu-Wing Mai, and Lin Ye. 1999. Resistance welding of carbon fibre reinforced thermoplastic composite using alternative heating element. *Composite Structures* 47 (1): 667–672.

44. Colak, Ziya Seyhan, Fazil Onder Sonmez, and Vahan Kalenderoglu. 2002. Process Modeling and optimization of resistance welding for thermoplastic composites. *Journal of Composite Materials* 36 (6): 721–744.
45. Chen, M., G. Zak, and P.J. Bates. 2015. Absorption coefficient measurement in laser transmission welding of thermoplastics. *International Polymer Processing* 30 (1): 38–43.
46. Bates, P.J., T.B. Okoro, and M. Chen. 2015. Thermal degradation of PC and PA6 during laser transmission welding. *Welding in the World* 59 (3): 381–390.
47. Gao, Jicheng, Chao Li, Unisha Shilpakar, and Yifu Shen. 2015. Improvements of mechanical properties in dissimilar joints of HDPE and ABS via carbon nanotubes during friction stir welding process. *Materials and Design* 86: 289–296.
48. Abbasi, M., B. Bagheri, and R. Keivani. 2015. Thermal analysis of friction stir welding process and investigation into affective parameters using simulation. *Journal of Mechanical Science and Technology* 29 (2): 861–866.
49. Rahmat, S.M., M. Hamdi, F. Yusof, and R. Moshwan. 2014. Preliminary study on the feasibility of friction stir welding in 7075 aluminium alloy and polycarbonate sheet. *Materials Research Innovations* 18 (S6): S6–515.
50. Rahbarpour, R., T. Azdast, H. Rahbarpour, and S.M. Shishavan. 2014. Feasibility study of friction stir welding of wood-plastic composites. *Science and Technology of Welding and Joining* 19 (8): 673–681.
51. Pirizadeh, Mehdi, Taher Azdast, Samrand Rash Ahmadi, Sajjad Mamaghani Shishavan, and Arvin Bagheri. 2014. Friction stir welding of thermoplastics using a newly designed tool. *Materials & Design* 54: 342–347.
52. Bilici, M.K., and A.I. Yükler. 2012. Influence of tool geometry and process parameters on macrostructure and static strength in friction stir spot welded polyethylene sheets. *Materials and Design* 33: 145–152.
53. Panneerselvam, K., and K. Lenin. 2014. Joining of Nylon 6 plate by friction stir welding process using threaded pin profile. *Materials and Design* 53: 302–307.
54. Ondruška, Michal, Marián Drienovský, Roman Čička, Milan Marônek, and Antonín Náplava. 2014. Optimizing the welding of plastics with the use of differential scanning calorimetry and thermogravimetric analysis. *Acta Polytechnica* 54 (3): 221–224.
55. Vacogne, C., and R. Wise. 2011. Joining of high performance carbon fibre/PEEK composites. *Science and Technology of Welding and Joining* 16 (4): 369–376.
56. Dubé, M., P. Hubert, A. Yousefpour, and J. Denault. 2007. Resistance welding of thermoplastic composites skin/stringer joints. *Composites Part A Applied Science and Manufacturing* 38 (12): 2541–2552.
57. Spancken, D., J. Decker, S. Ruotsalainen, P. Laakso, and A. Büter. 2015. Fatigue design of thermoplastic laser welds. *Welding in the World* 59 (1): 65–70.
58. Rauschenberger, J., D. Vogler, C. Raab, and U. Gubler. 2015. Diffractive beam shaping for enhanced laser polymer welding. In *SPIE LASE*, 935110–935110. International Society for Optics and Photonics.
59. Zhong, Xue Jiao, Cai Lian Fan, Hui Xia Liu, Pin Li, and Xiao Wang. 2015. Light scattering of HDPE and LDPE in laser transmission welding. In *Key engineering materials*, vol. 667, 95–101. Trans Tech Publications.
60. Amanat, Negin, Cedric Chaminade, John Grace, David R. McKenzie, and Natalie L. James. 2010. Transmission laser welding of amorphous and semi-crystalline poly-ether–ether–ketone for applications in the medical device industry. *Materials and Design* 31 (10): 4823–4830.
61. Speka, Maryna, Simone Matteï, Michel Pilloz, and Mariana Ilie. 2008. The infrared thermography control of the laser welding of amorphous polymers. *NDT and E International* 41 (3): 178–183.
62. Volpe, Annalisa, Francesca Di Niso, Caterina Gaudiuso, Andrea De Rosa, Rebeca Martìnez Vàzquez, Antonio Ancona, Pietro Mario Lugarà, and Roberto Osellame. 2015. Femtosecond fiber laser welding of PMMA. In *SPIE LASE*, 935106–935106. International Society for Optics and Photonics.

63. Jakobsen, Tom B., Roderic C. Don, and John W. Gillespie. 1989. Two dimensional thermal analysis of resistance welded thermoplastic composites. *Polymer Engineering & Science* 29 (23): 1722–1729.
64. Pitchumani, R., S. Ranganathan, R.C. Don, J.W. Gillespie, and M.A. Lamontia. 1996. Analysis of transport phenomena governing interfacial bonding and void dynamics during thermoplastic tow-placement. *International Journal of Heat and Mass Transfer* 39 (9): 1883–1897.
65. Xiao, X.R., S.V. Hoa, and K.N. Street. 1992. Processing and modelling of resistance welding of APC-2 composite. *Journal of Composite Materials* 26 (7): 1031–1049.
66. Wang, C.Y., P.J. Bates, and G. Zak. 2009. Optical properties characterization of thermoplastics used in laser transmission welding: Transmittance and reflectance. In *Proceedings of the SPE ANTEC*, 1278–1282.
67. Aden, M., A.M. Roesner, and A. Olowinsky. 2010. Optical characterization of polycarbonate: Influence of additives on optical properties. *Journal of Polymer Science Part B: Polymer Physics* 48: 451–455.
68. Coelho, J.M.P., M.A. Abreu, F.C. Rodrigues. 2004. Methodologies for determining thermoplastic films optical parameters at 10.6 mm laser wavelength. *Polymer Testing* 23: 307–312.
69. Kagan, V.A., R.G. Bray, and W.P. Kuhn. 2002. Laser transmission welding of semi-crystalline thermoplastics—Part I: Optical characterization of nylon based plastics. *Journal of Reinforced Plastics and Composites* 21 (12): 1101–1122.
70. Rhew, M., A. Mokhtarzadeh, A. Benatar. 2003. Diode laser characterization and measurement of optical properties of polycarbonate and high-density polyethylene. In *Proceedings of the SPE ANTEC*, 1056–1060.
71. Dosser, L., K. Hix, K. Hartke, R. Vaia, and M. Li. 2004. Transmission welding of carbon nanocomposites with direct-diode and Nd:YAG solid state lasers. *Proceedings of SPIE* 5339: 465–474.
72. Chen, M.L., G. Zak, and P.J. Bates. 2011. Effect of carbon black on light transmission in laser welding of thermoplastics. *Journal of Materials Processing Technology* 211: 43–47.
73. Azhikannickal, E., P.J. Bates, G. Zak. 2012. Laser light transmission through thermoplastics as a function of thickness and laser incidence angle: Experimental and modelling. *ASME Journal of Manufacturing Science and Engineering* 134: 061007-1–6.
74. Azhikannickal, E., P.J. Bates, and G. Zak. 2012. Thermal imaging technique to characterize laser light reflection from thermoplastics. *Optics & Laser Technology* 44: 1456–1462.
75. Xu, Xin Feng, Philip J. Bates, and Gene Zak. 2015. Effect of glass fiber and crystallinity on light transmission during laser transmission welding of thermoplastics. *Optics & Laser Technology* 69: 133–139.
76. Voisiat, B., D. Gaponov, P. Gečys, L. Lavoute, M. Silva, A. Hideur, N. Ducros, and G. Račiukaitis. 2015. Material processing with ultra-short pulse lasers working in 2 μm wavelength range. In *SPIE LASE*, 935014–935014. International Society for Optics and Photonics.
77. Visco, A.M., G. Galtieri, L. Torrisi, and C. Scolaro. 2015. Properties of single and double lap polymeric joints welded by a diode laser. *International Journal of Polymer Analysis and Characterization*.
78. Van der Straeten, Kira, Christoph Engelmann, Alexander Olowinsky, and Arnold Gillner. 2015. Laser transmission welding of long glass fiber reinforced thermoplastics. In *SPIE LASE*, 93560H–93560H. International Society for Optics and Photonics.
79. Tamrin, K.F., Y. Nukman, and N.A. Sheikh. 2015. Laser spot welding of thermoplastic and ceramic: An experimental investigation. *Materials and Manufacturing Processes*.
80. Ruotsalainen, Saara, Petri Laakso, and Veli Kujanpää. 2015. Laser welding of transparent polymers by using quasi-simultaneous beam off-setting scanning technique. *Physics Procedia* 78: 272–284.
81. Rodríguez-Vidal, E., C. Sanz, C. Soriano, J. Leunda, and G. Verhaeghe. 2015. Effect of metal micro-structuring on the mechanical behavior of polymer-metal laser T-joints. *Journal of Materials Processing Technology*.
82. Rauschenberger, J., A. Cenigaonaindia, J. Keseberg, D. Vogler, U. Gubler, and Fernando Liébana. 2015. Laser hybrid joining of plastic and metal components for lightweight components. In *SPIE LASE*, 93560B–93560B. International Society for Optics and Photonics.

83. Wang, C.Y., P.J. Bates, and G. Zak. 2010. Optical properties characterization of thermoplastics used in laser transmission welding: Scattering and absorbance. *Advanced Materials Research* 97–101: 3836–3841.
84. Geiger, M., T. Frick, and M. Schmidt. 2009. Optical properties of plastics and their role for the modelling of the laser transmission welding process. *Production Engineering Research and Development* 3 (1): 49–55.
85. Lu, H.M., and A. Benatar. 1991. Sequential ultrasonic welding of PEEK/Graphite composite parts. In *Proceedings of the annual technical conference ANTEC*. Montreal.
86. Woizeschke, Peer, Eugen Mosgowoi, and Frank Vollertsen. 2015. Decreasing pore formation in multiple-sheet laser joining with interfacial polymeric contaminations. *Welding in the World* 1–10.
87. Visco, A.M., V. Brancato, L. Torrisi, and M. Cutroneo. 2014. Employment of Carbon Nanomaterials for Welding Polyethylene Joints with a Nd: YAG Laser. *International Journal of Polymer Analysis and Characterization* 19 (6): 489–499.
88. Tan, Xianghu, Jing Zhang, Jiguo Shan, Shanglu Yang, and Jialie Ren. 2015. Characteristics and formation mechanism of porosities in CFRP during laser joining of CFRP and steel. *Composites Part B Engineering* 70: 35–43.
89. Jiang, Xin, Soni Chandrasekar, and Changhai Wang. 2015. A laser microwelding method for assembly of polymer based microfluidic devices. *Optics and Lasers in Engineering* 66: 98–104.
90. Cosson, Benoit, Mylène Deléglise, and Wolfgang Knapp. 2015. Numerical analysis of thermoplastic composites laser welding using ray tracing method. *Composites Part B: Engineering* 68: 85–91.
91. Berger, Stefan, Florian Oefele, and Michael Schmidt. 2015. Laser transmission welding of carbon fiber reinforced thermoplastic using filler material—A fundamental study. *Journal of Laser Applications* 27 (S2): S29009.
92. Asséko, André Chateau Akué, Benoît Cosson, Fabrice Schmidt, Yannick Le Maoult, and Eric Lafranche. 2015. Laser transmission welding of composites-Part A: Thermo-physical and optical characterization of materials. *Infrared Physics & Technology*.
93. Acherjee, Bappa, Arunanshu S. Kuar, Souren Mitra, and Dipten Misra. 2015. Empirical modeling and multi-response optimization of laser transmission welding of polycarbonate to ABS. *Lasers in Manufacturing and Materials Processing* 1–21.
94. Wang, Xiao, Hao Chen, and Huixia Liu. 2014. Investigation of the relationships of process parameters, molten pool geometry and shear strength in laser transmission welding of polyethylene terephthalate and polypropylene. *Materials & Design* 55: 343–352.
95. Rodríguez-Vidal, E., I. Quintana, and C. Gadea. 2014. Laser transmission welding of ABS: Effect of CNTs concentration and process parameters on material integrity and weld formation. *Optics & Laser Technology* 57: 194–201.
96. Acherjee, Bappa, Arunanshu S. Kuar, Souren Mitra, and Dipten Misra. 2014. Laser transmission welding of polycarbonates: Experiments, modeling, and sensitivity analysis. *The International Journal of Advanced Manufacturing Technology* 78 (5–8): 853–861.
97. Yusof, Farazila, Miyashita Yukio, Mutoh Yoshiharu, and Mohd Hamdi Abdul Shukor. 2012. Effect of anodizing on pulsed Nd: YAG laser joining of polyethylene terephthalate (PET) and aluminium alloy (A5052). *Materials and Design* 37: 410–415.
98. Rodríguez-Vidal, Eva, Iban Quintana, Jon Etxarri, Urko Azkorbebeitia, Deitze Otaduy, Francisco González, and Fernando Moreno. 2012. Optical design and development of a fiber coupled high-power diode laser system for laser transmission welding of plastics. *Optical Engineering* 51 (12): 124301–124301.
99. Acherjee, Bappa, Subrata Mondal, Bipan Tudu, and Dipten Misra. 2011. Application of artificial neural network for predicting weld quality in laser transmission welding of thermoplastics. *Applied Soft Computing* 11 (2): 2548–2555.
100. Acherjee, Bappa, Arunanshu S. Kuar, Souren Mitra, and Dipten Misra. 2011. Application of grey-based Taguchi method for simultaneous optimization of multiple quality characteristics in laser transmission welding process of thermoplastics. *The International Journal of Advanced Manufacturing Technology* 56 (9–12): 995–1006.

101. Khodabakhshi, F., M. Haghshenas, S. Sahraeinejad, J. Chen, B. Shalchi, J. Li, and A.P. Gerlich. 2014. Microstructure-property characterization of a friction-stir welded joint between AA5059 aluminum alloy and high density polyethylene. *Materials Characterization* 98: 73–82.
102. Gao, Jicheng, Yifu Shen, Jingqing Zhang, and Haisheng Xu. 2014. Submerged friction stir weld of polyethylene sheets. *Journal of Applied Polymer Science* 131 (22).
103. De Saracibar, C. Agelet, Michele Chiumenti, Miguel Cervera, N. Dialami, and Anthony Seret. 2014. Computational modeling and sub-grid scale stabilization of incompressibility and convection in the numerical simulation of friction stir welding processes. *Archives of Computational Methods in Engineering* 21 (1): 3–37.
104. Azarsa, Ehsan, and Amir Mostafapour. 2014. Experimental investigation on flexural behavior of friction stir welded high density polyethylene sheets. *Journal of Manufacturing Processes* 16 (1): 149–155.
105. Ahmadi, H., N.B. Mostafa Arab, and F. Ashenai Ghasemi. 2014. Optimization of process parameters for friction stir lap welding of carbon fibre reinforced thermoplastic composites by Taguchi method. *Journal of Mechanical Science and Technology* 28 (1): 279–284.
106. Benatar, A., and T.G. Gutowski. 1986. Methods for fusion bonding thermoplastic composites. *SAMPE Quart* 18 (1): 35–42.
107. Dialami, Narges, Michele Chiumenti, Miguel Cervera, and Carlos Agelet de Saracibar. 2013. An apropos kinematic framework for the numerical modeling of friction stir welding. *Computers & Structures* 117: 48–57.
108. Dashatan, Saeid Hoseinpour, Taher Azdast, Samrand Rash Ahmadi, and Arvin Bagheri. 2013. Friction stir spot welding of dissimilar polymethyl methacrylate and acrylonitrile butadiene styrene sheets. *Materials & Design* 45: 135–141.
109. Bagheri, Arvin, Taher Azdast, and Ali Doniavi. 2013. An experimental study on mechanical properties of friction stir welded ABS sheets. *Materials and Design* 43: 402–409.
110. Yusof, F., Y. Miyashita, N. Seo, Y. Mutoh, and R. Moshwan. 2012. Utilising friction spot joining for dissimilar joint between aluminium alloy (A5052) and polyethylene terephthalate. *Science and Technology of Welding and Joining* 17 (7): 544–549.
111. Ratanathavorn, Wallop. 2012. Hybrid joining of aluminum to thermoplastics with friction stir welding.
112. Kiss, Z., and T. Czigány. 2012. Effect of welding parameters on the heat affected zone and the mechanical properties of friction stir welded poly (ethylene-terephthalate-glycol). *Journal of Applied Polymer Science* 125 (3): 2231–2238.
113. Kiss, Zoltan, and Tibor Czigany. 2012. Microscopic analysis of the morphology of seams in friction stir welded polypropylene. *Express Polymer Letters* 6 (1): 54–62.
114. Bozkurt, Yahya. 2012. The optimization of friction stir welding process parameters to achieve maximum tensile strength in polyethylene sheets. *Materials and Design* 35: 440–445.
115. Bilici, Mustafa Kemal, and Ahmet Irfan Yükler. 2012. Influence of tool geometry and process parameters on macrostructure and static strength in friction stir spot welded polyethylene sheets. *Materials & Design* 33: 145–152.
116. Bilici, Mustafa K. 2012. Effect of tool geometry on friction stir spot welding of polypropylene sheets. *Express Polymer Letters* 6 (10): 805–813.
117. Azarsa, Ehsan, Amir Mostafapour Asl, and Vahid Tavakolkhah. 2012. Effect of process parameters and tool coating on mechanical properties and microstructure of heat assisted friction stir welded polyethylene sheets. In *Advanced materials research*, vol. 445, 765–770.
118. Ahmadi, Hedi, Nasrollah Bani Mostafa Arab, and Faramarz Ashenai Ghasemi. 2012. Application of Taguchi method to optimize friction stir welding parameters for polypropylene composite lap joints. Archives Des Sciences 65 (7).
119. Zhang, De Fen, Fei Long, Xiao Wen Chen, Xiang Qian Wen, and Hong Song Luo. 2011. Review on research status of friction stir welding technology. In *Advanced Materials Research*, vol. 335, 379–382.
120. Saeedy, S., and M.K. Besharati Givi. 2011. Investigation of the effects of critical process parameters of friction stir welding of polyethylene. *Proceedings of the Institution of Mechanical Engineers, Part B: Journal of Engineering Manufacture* 09544054JEM1989.

121. Rai, R., A. De, H.K.D.H. Bhadeshia, and T. DebRoy. 2011. Review: Friction stir welding tools. *Science and Technology of Welding and Joining* 16 (4): 325–342.
122. Payganeh, G. H., N.B. Mostafa Arab, Y. Dadgar Asl, F.A. Ghasemi, and M. Saeidi Borou-jeni. 2011. Effects of friction stir welding process parameters on appearance and strength of polypropylene composite welds. *International Journal of Physical Sciences* 6 (19): 4595–4601.
123. Mert, S., and A. Arici. 2011. Design of optimal joining for friction stir spot welding of polypropylene sheets. *Science and Technology of Welding and Joining* 16 (6): 522–527.
124. Amancio-Filho, S.T., C. Bueno, J.F. Dos Santos, N. Huber, and E. Hage. 2011. On the feasibility of friction spot joining in magnesium/fiber-reinforced polymer composite hybrid structures. *Materials Science and Engineering A* 528 (10): 3841–3848.
125. Bilici, Mustafa K. 2012. Effect of tool geometry on friction stir spot welding of polypropylene sheets. *Express Polym Lett* 6 (10): 805–813.
126. Sadeghian, Nasser, and Mohammad Kazem Besharati Givi. 2015. Experimental optimization of the mechanical properties of friction stir welded acrylonitrile butadiene styrene sheets. *Materials and Design* 67: 145–153.
127. Gonçalves, J., J.F. dos Santos, L.B. Canto, and S.T. Amancio-Filho. 2015. Friction spot welding of carbon fiber-reinforced polyamide 66 laminate. *Materials Letters* 159: 506–509.
128. Vijendra, Bandari, and Abhay Sharma. 2015. Induction heated tool assisted friction-stir welding (i-FSW): A novel hybrid process for joining of thermoplastics. *Journal of Manufacturing Processes* 20: 234–244.
129. Hoseinlaghab, Sadegh, Seyed Sajad Mirjavadi, Nasser Sadeghian, Iraj Jalili, M. Azarbarmas, and Mohammad Kazem Besharati Givi. 2015. Influences of welding parameters on the quality and creep properties of friction stir welded polyethylene plates. *Materials & Design* 67: 369–378.
130. Liu, F.C., J. Liao, and K. Nakata. 2014. Joining of metal to plastic using friction lap welding. *Materials and Design* 54: 236–244.
131. Lambiase, F., A. Paoletti, and A. Di Ilio. 2015. Mechanical behaviour of friction stir spot welds of polycarbonate sheets. *The International Journal of Advanced Manufacturing Technology* 1–14.
132. Mendes, N., A. Loureiro, C. Martins, P. Neto, and J.N. Pires. 2014. Morphology and strength of acrylonitrile butadiene styrene welds performed by robotic friction stir welding. *Materials and Design* 64: 81–90.

Chapter 3
Ultrasonic Welding of Polymers

3.1 Introduction to USW

Ultrasonic welding of polymers is an economically viable joining technique which eludes the addition of solvents or adhesives but causes localized heating at the interface by using relatively short cycle times [1]. Ultrasound welding is a high-frequency continuous joining technique. It is appropriate for joining small and larger regions in a sequential manner [2]. It is the technique that offers a good alternative to automotive, medical, packaging, appliance, textile, electronics, and others. The merits of ultrasonically assembled parts reveal clean and reliable bonds to the components. Ultrasonic welding is a swift process wherein a frequency in the range of 18–70 kHz is being used. Output value differs from hundreds of watts to a few kilowatts. Ultrasonic welding machine comprises a booster, transducer, horn, and electrical power supply (generator). The transducer (process controller) receives the electrical signal supply from the generator. Transducer converts electrical signal into mechanical longitudinal vibrations which are transmitted to horn to bond the interface of the material.

Scientific literature reports reveal that the heat source in ultrasonic welding of polymers is governed by the hysteresis losses. The problem of heating of plastics at the stage of cyclic deformation is resolved. Thermal degradation is prevented in the bonding of thermoplastics and thermoset composites while setting the heating time under 1 s. This ultrasonic welding yields ultra-fast efficient welds within a short duration with dissimilar material mixture [3].

Energy directors are used to concentrate the heat at the weld interfaces during ultrasonic welding. This energy director acts as alternative energy solution for joining thermoplastic composites depending on the application. Analysis of dissipated power, welding energy, and time are considered during dissimilar material welding using ultrasonic technique. Energy director is used to simplify the joining of thermoplastics with no compromise on the quality of the weld [4]. Besides, it is emphasized

S. A. Vendan et al., *Confluence of Multidisciplinary Sciences for Polymer Joining*,
https://doi.org/10.1007/978-981-13-0626-6_3

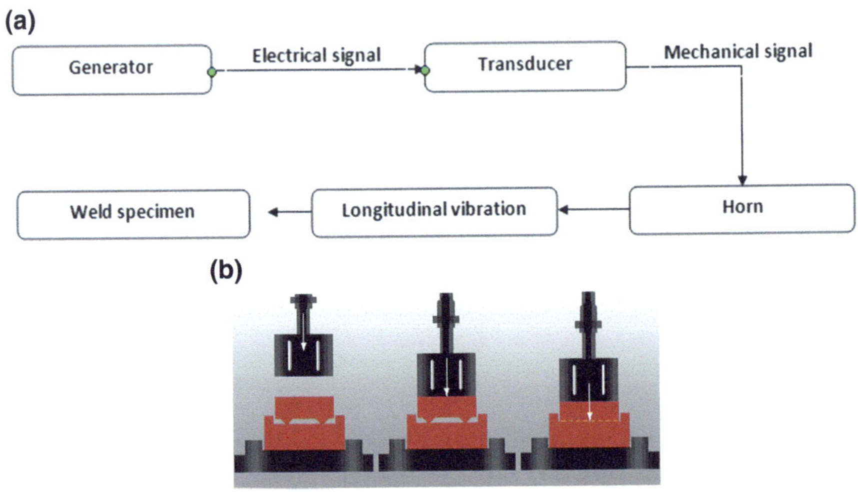

Fig. 3.1 **a** Schematic of ultrasonic welding process. **b** Ultrasonic welding processes of plastics [2]

that inclusion of a plasticizer has a strong influence on the processing window and on the weld quality [5].

The waterproof composites are welded using ultrasonic welding process for the specialized applications in the studies. Various recommendations are made to enhance the ultrasonic welding technique to offer theoretical guidance for further research and usages in different industries [5]. Production of micro-devices made from thermoplastic materials using ultrasonic welding method is an optimal technology for the current manufacturing scenario. For small-scale production as in case of micro-devices industry, it is less expensive as compared to other methods of plastic joining. Ultrasonic welding process is being deployed for past ten years and adapted in the industry for multiple purposes as shown in Fig. 3.1a, b [2–6].

The properties of the ultrasonically welded specimens are altered due to inclusion of additives during production owing to environmental influences [7]. In the ultrasonic welding, thin metal wires are used as filler materials and placed among sidewall and lid foil. After turning on the ultrasound, friction heat is produced among wire and plastics. The polymer is melted with the aid of filler material along the energy director. Subsequently, the wire solidifies and gets embedded in Fig. 3.2.

3.1.1 Horn Profile Analysis

The conclusions observed from Table 3.1 are presented as follows:

1. Nodal region is developed with higher temperature in the aluminum horns as compared to other horn materials. The temperature rise reduces the resonant

Table 3.1 Analysis of various horn profiles [8]

S. no.	Horn material	Displacement amplitude (μm)	Alternating stress at the nodal region (MPa)	Cumulative damage (ratio of cycles used/allowed)	Predicted safe working temperature (°C)	ANSYS design for horn profile
1	Aluminum	24	48.6	0.10	105.0	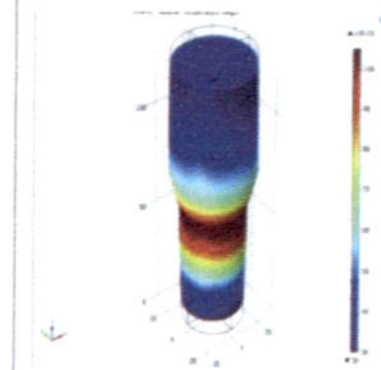
		36	73.0	0.10	198.8	
		44	89.3	0.85 (<1)	282.2	
2	Titanium	24	86.7	0.10	60.1	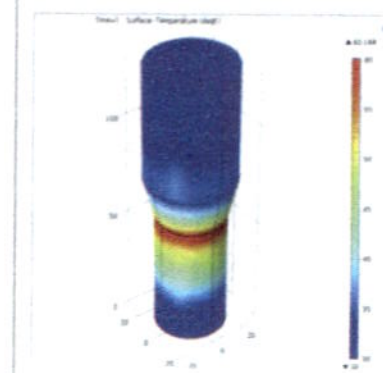
		44	159.0	0.10	131.4	
		77	278.3	0.82 (<1)	340.5	

(continued)

Table 3.1 (continued)

S. no.	Horn material	Displacement amplitude (µm)	Alternating stress at the nodal region (MPa)	Cumulative damage (ratio of cycles used/allowed)	Predicted safe working temperature (°C)	ANSYS design for horn profile
3	Mild steel	24	138.5	0.10	85.4	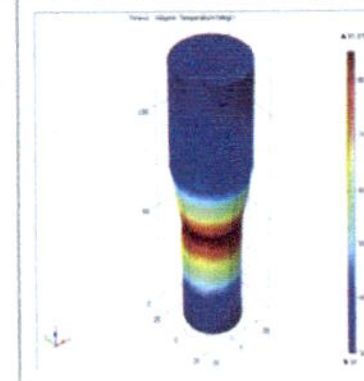
		44	254.0	0.10	116.6	
		47	271.3	1.00 (<1)	242.7	
4.	Stainless steel	24	122.3	0.10	76.0	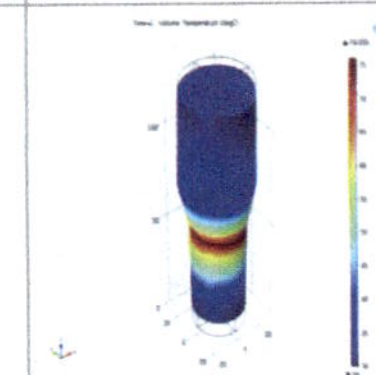
		30	153.7	0.10	101.9	
		37	189.5	0.90 (<0)	139.5	

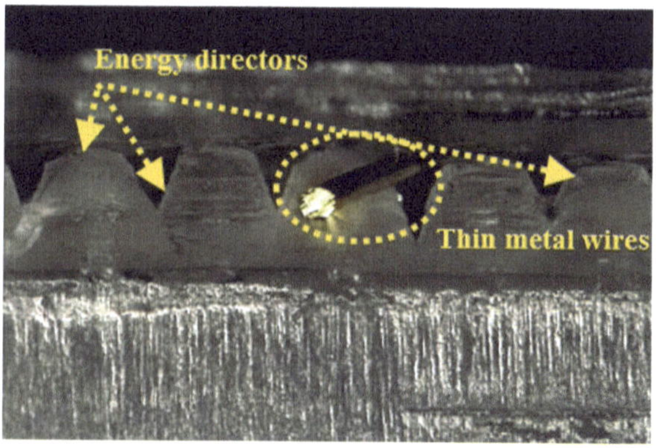

Fig. 3.2 Thin metal wires acted as energy director [7]

frequency of the horn while diminishing its performance. It may be established that aluminum horns are inappropriate for mass fabrication in the industries.

2. Titanium horns experiences lesser temperature increase among the horns made of other materials because of its lower internal damping coefficient, and henceforth, the resonant frequency remains stable.
3. The mild steel horn is suitable for operating at higher temperatures compared to other steel horns. The parts joined with mild steel horn yields higher strength compared to other steel horns.
4. Titanium horns yield higher weld strength as compared to the other horns for ABS plastic components joining while titanium horn improved the energy utilization with negligible acoustic losses.
5. Analysis of the cyclic loads is observed that the highest displacement amplitude for failure of the titanium horn is found at 77 μm with 106 cycles with maximum operation period for the titanium horn [8].

An ultrasonic tool (horn) provides the energy to the specimen to weld in the nodal region produced on either side of the joint region. Two rigid mount boosters are utilized, each coupled to both side of the joint zone. One of the rigid mount boosters has been attached with a transducer. Housing assembly of the ultrasonic machine has been attached with two ultrasonic stack mounting rings at a non-nodal zone and attached with one of the rigid mount boosters at the nodal zone [9].

Levy et al. [10] discussed the probable mechanisms of interfacial adhesion among thermoplastic and thermoset plastics while also presenting in detail the investigated results on the possible approaches to enhance interfacial adhesion for the intentions of bonding fiber-reinforced polymer (FRP) composite domains by fusion welding. In the ultrasonic systems, designing large horns poses a challenge which involves configuration to longitudinal mode at the working frequency with uniform displacement amplitude at the surface of the horn. To limit the modal behavior, large horns are generally modeled with slots parallel to the longitudinal motion. To attain the required longitudinal mode with even amplitude at the horn surface, the dimensions and the position of the slots are complex [11]. Regression equation and the plots of main impacts of the slot factors are considered while choosing the optimum design factors. The horn produced based on this design is tuned and tested successfully in the industries [12]. Pressure and surface roughness among the closely contacted two plates of the polymer are the deciding parameters for the polymer characteristics [13].

3.1.2 Healing Effect

Nonhof and Luiten [14] investigated to define the healing degree D_h (Eq. 3.3) as the ratio among the ultimate interfacial bond strength and the instantaneous interfacial bond strength. The model for healing degree is suggested by Nonhof with parameters such as welding time of the full interface bond strength (Eq. 3.1) achieved by Eq. (3.1):

$$\frac{\partial D_h}{\partial t} = \left(\frac{1}{t_w(T)}\right)^{\frac{1}{4}} \tag{3.1}$$

For low molecular weight of polymers, this welding time is same as the repetition time, but the response is converse for high molecular weight of the polymer (full strength is attained before a full repetition of the macromolecule). The welding time (Eq. 3.2) depends on the temperature T typically modeled using the Arrhenius law [15]:

$$t_w = A_{erp} \exp\left(\frac{E_a}{RT}\right) \tag{3.2}$$

A_{rep} represents the pre-exponential coefficient based on the material. The following equation presents the healing computed with Eq. (3.3)

$$D_h = \int_{t_0}^{t} \frac{1}{t_w(T)^{\frac{1}{4}}} dt \tag{3.3}$$

Two phases in the welding process: (i) a phase of localized heating due to the tip effect till it reaches an adequate drop of mechanical properties followed by (ii) a flow phase where a fold of polymer is produced and moves forward till filling the gap among the plates and eventually attaining adhesion. Energy director squeezes for higher holding time experiencing higher temperature that leads to poor welding [16].

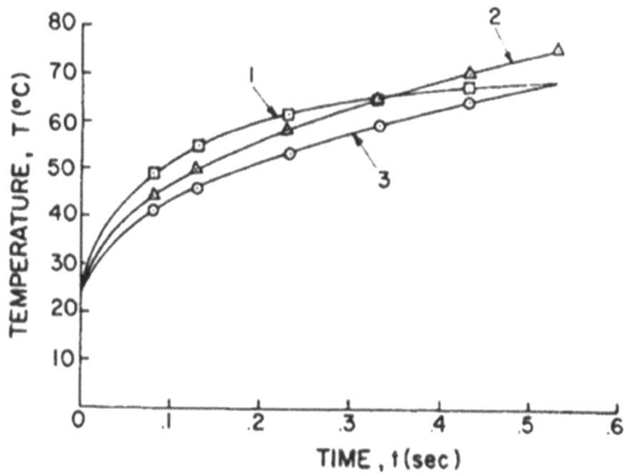

Fig. 3.3 Typical temperature rise results from experiments: 1. Actual response; 2. Theoretical response where thermal flux is calculated from the temperature rise at 0.33 s 3. Theoretical response where thermal flux is calculated from the temperature rise at 0.53 s [17]

3.2 Prominent Research Reports on USW

3.2.1 *Evolution of the Ultrasonic Welding During the Years of 1980s*

Primary features of ultrasonic welding of thermoplastics are discussed by Frankel and Wang. Establishment of simple model is carried out by characterizing the increase in temperature at the joint until the glass transition temperature is achieved. The temperature rises more swiftly and is directly proportional to weld time beyond this point (Fig. 3.3). The increased amplitude of vibration rises in the rate of increase of temperature. Dimension analysis technique is espoused to develop the correlation among interface temperature and weld strength. The observation indicates that the method can be optimized with respect to the weld strength by monitoring the power input. High weld strength is achievable with the application of an optimal load. Application of energy is less to a certain extent for the overall efficiency of the process [17].

Carboni [18] reported the investigations on heating and bonding processes in ultrasonic welding. Temperatures at the weld interface are recorded for the polystyrene samples with various welding conditions during the welding process. The other parameters (amplitude of vibration, power input, and amount of deformation) are measured during welding process. Initially, higher rate of heating at the interface is observed during welding till temperature reaches about 250 °C, but declines thereon at the interface. Higher temperature at the interface is reached prior to joining of the samples. The samples are joined at glass transition temperature. Destructive test is carried out to determine the mechanical strength of the welded samples using special machine which is customized by combining with compressional and torsional loads. From the test results, the amount of energy and rate of material flowing out of the interface are influential parameters to the weld strength. Single disk and double disk

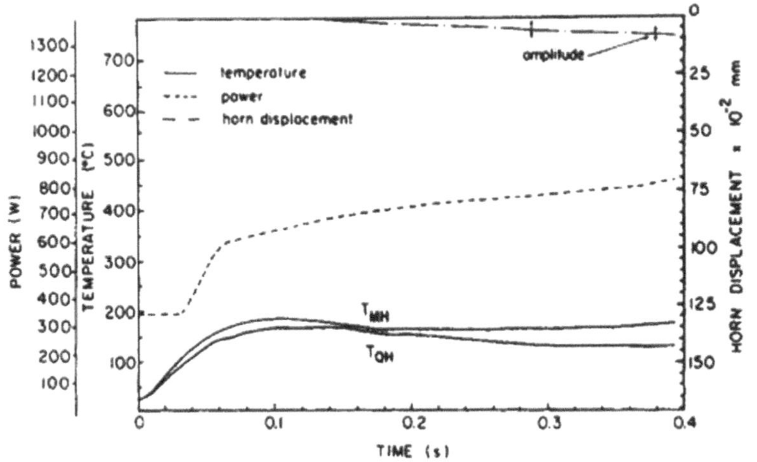

Fig. 3.4 Temperature, power, and sonotrode displacement traces for a single disk experiment [18]

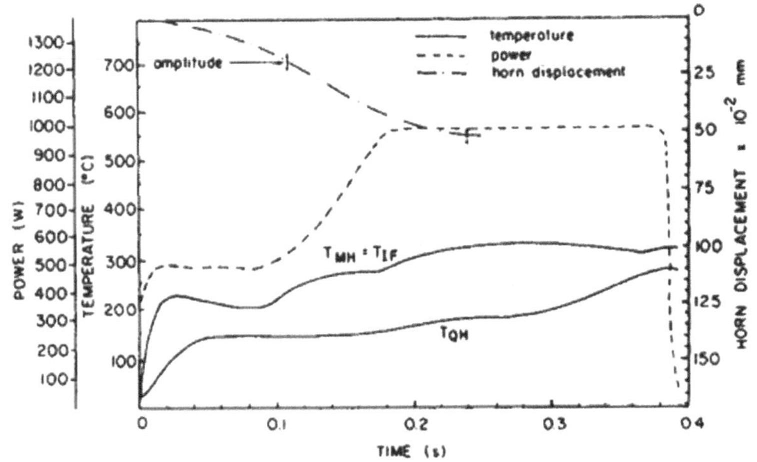

Fig. 3.5 Temperature, power, and sonotrode displacement traces for a double disk experiment [18]

experiments are performed, and the results are provided for the temperature, power, and sonotrode as shown in Figs. 3.4 and 3.5, respectively.

Suresh et al. [19] made assessment of the simple model rod, sound field effects, energy conversion, and energy transmission and reported the responses. The influence of ultrasonic energy transmission and energy conversion in the geometries (energy director) of the samples is examined (Fig. 3.6). Evaluation of welding capacity of thermoplastics is set with certain criteria in the end. It is observed that the weld quality is majorly affected by the joining pressure.

Tsujino et al. [20] carried out an investigation with far-field (the distance among the ultrasonic horn and the joint is greater than 6 mm) ultrasonic welding of amorphous

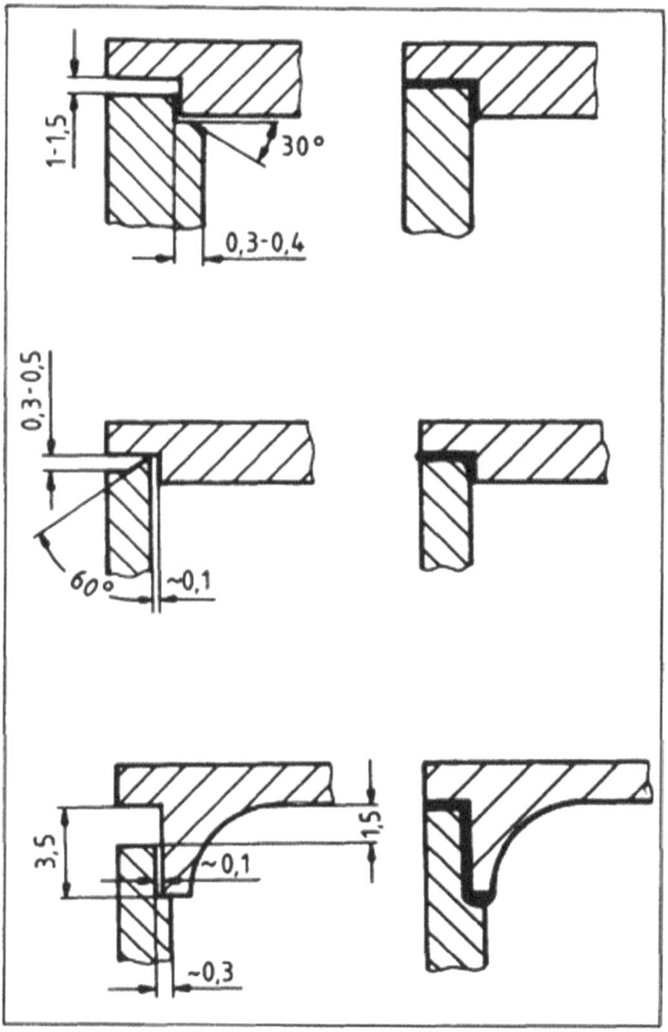

Fig. 3.6 Contact surface geometries [19]

(polystyrene and acrylo butadiene styrene) and semicrystalline (polypropylene and polyethylene) polymers. Amorphous polymers are joined successfully using far-field welding. Joint interface is seen with improved weld strength by increasing the amplitude of vibration. Higher weld strength is also achieved by rising the weld time and/or weld pressure. Semicrystalline polymers are not that successfully joined using far-field ultrasonic welding. No melting or little melting at the joint interface is observed with little deformation at the part–horn interface. Prediction of vibration amplitude occurring at the joint interface is performed through a model for wave propagation in viscoelastic materials. The model signified that enhancement of far-field welding of semicrystalline polymers is carried out by increasing the length of the specimens to a half wavelength and by amplifying the amplitude of vibration at the joint interface.

Models such as mechanics and vibration of the parts, heat transfer, intermolecular diffusion, flow and wetting, and viscoelastic heating are used for joining of composites through the ultrasonic welding process. Experimentation is carried out to validate the model to predict the melting and flow pattern in welding process. The model also specifies the feasibility of supervising joint quality by assessing the dynamic mechanical impedance of the parts throughout welding, which is also confirmed experimentally such as supervising the magnitude of the impedance. The dynamic impedance of the interface of composites is illustrated to increase quickly when the melt fronts of the energy directors join at the end of process. The ultrasonic welding of thermoplastic composites uses the closed-loop control procedures which are established from the influence of dynamic impedance. Exceptional strength of joint is achieved for the polyether ether ketone (PEEK) graphite APC-2 composites through ultrasonic welding [21].

Multiple benefits are available when ultrasonic welding technique is employed for joining thermoplastics. Ultrasonic welding is economical, swift, hazard-free, and easily automated. Near-field ultrasonic welding of amorphous (polystyrene and acrylonitrile butadiene styrene) and semicrystalline (polypropylene and polyethylene) polymers is utilized in this investigation which is the distance prevailing between the sonotrode and the joint interface is 6 mm or less [22]. The assessment of the dynamic mechanical moduli of the polymers is carried out for high-frequency ultrasonic wave propagation and shrinkage measurements. Lumped parameter model is used to foresee energy dissipation and heating rates [23]. It is observed from the experimental results that differences in the welding pressure had minimum influence on joint strength or energy dissipation; magnifying the amplitude of vibration produced higher weld strength and energy dissipation. Increasing the weld time (>1.5 s) enhanced strength where strength of semicrystalline polymers leveled with the strength of amorphous polymers. Higher weld strength is achieved with increased weld time, i.e., 0.8 s (high power and load are required) for the amorphous polymers. Weld quality is assessed from the supervision of the static displacement or collapse and energy dissipation [24].

3.2.2 Progresses in the Ultrasonic Welding During the Years of 1990s

The features of glass fiber-reinforced thermoplastic (GFRTP) are maximum affected through the category of matrix, category, or geometry of the fiber and technique of reinforcement. GFRTP is generally used due to its broad range of elastic deformation and its strength, where second substitute is superior than metal. An efficient welding technique other than traditional mechanical fastening and joining techniques is

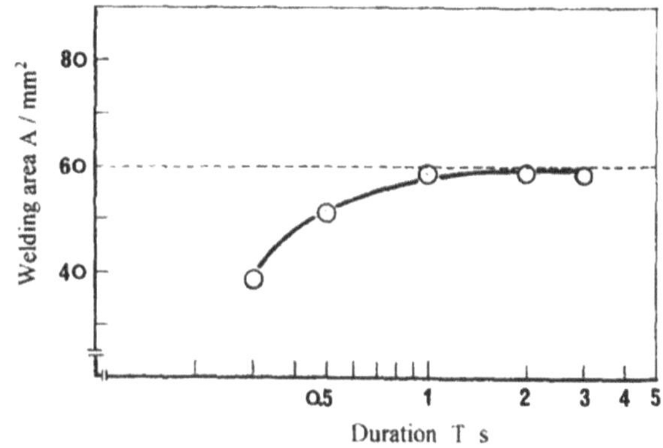

Fig. 3.7 Effects of the duration, T, on the joint region, A, of ultrasonically joined products [25]

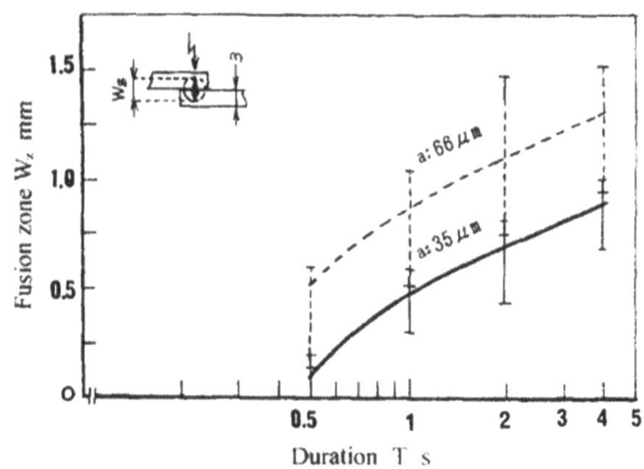

Fig. 3.8 Relationship between fusion condition of welded surface, W, and duration, T [25]

found for joining GFRTP which has wide usages. In this investigation, ultrasonic welding technique is efficiently used for joining of sheets (GFRTP). Major findings are discussed feasibilities of ultrasonic welding, features of the weld strength, and workability of the joined components in this investigation. Two types of fiber-reinforced plastic made with different fabrication methods are compared. Effects of duration (T) on the joint region of ultrasonically joined products are shown in Fig. 3.7. And also relationship among fusion condition of welded surface and weld region is illustrated in Fig. 3.8 [25].

Benatar and Cheng [26] discussed the PEEK–carbon composites joined using ultrasonic welding to better realize the method and identify the optimum welding parameters. Variation in the parameters is observed for the welding time and applied pressure. The horn to specimen interface is applied with optimum pressure which is 3.8 MPa. Fracture tests with various modes such as opening and shear are conducted to assess the joint properties. From the observation, it is determined that the optimum welding time highly relies on the geometry of the specimen being joined; this influences the efficiency of conversion of the ultrasound to thermal energy in the PEEK–carbon composite. On the other hand, for both categories of sample eval-

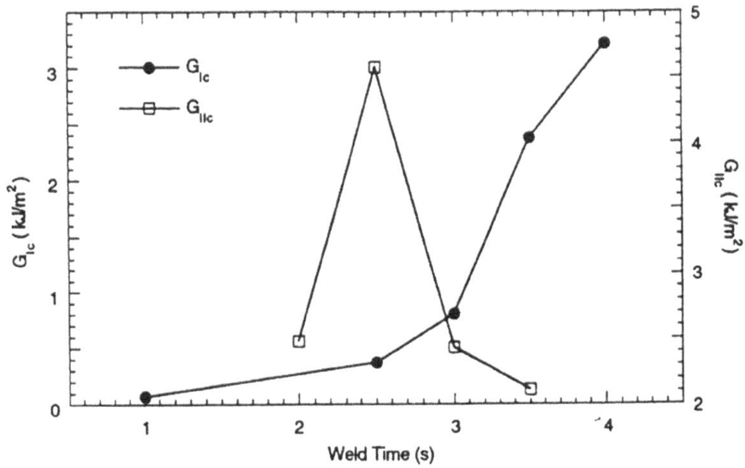

Fig. 3.9 Difference of critical strain energy release rate as a function of weld time for mode-1 and mode-2 testing [26]

uated (mode-1 and mode-2) the optimum weld strength is obtained with respect to a specific value of total energy input (6.8 J/mm^2). And this provides critical strain energy release rates (G,c = 3.2 kJ/m^2 and $G_{,1.}$ = 4.6 k/m^2). The difference in the energy (Fig. 3.9) is precisely 10% in both sides produced in reduction of properties by about half its values. Relationship between weld energy and weld time for mode-1 and mode-2 specimens is shown in Fig. 3.10. Difference of critical strain energy release rate for mode-1 and mode-2 as a function of weld energy input is illustrated in Fig. 3.11. Therefore, the weld energy can be assumed as a reliable regulating parameter to create welds with strength to those of molded product.

Benatar and Gutowski [27] reported the direct welding features of an ultrasonic plastic welding with vibration frequencies of longitudinal horn (maximum 90 kHz) and torsional vibration horn (minimum 20 kHz) with small and large vibrational amplitude. Welding features are enhanced by the usage of small and large vibration amplitude with high and low vibration frequencies which produced minimum damage to the samples because of vibration resulting in traditional systems. The two-vibration-system welding machine components are upper longitudinal vibration source (90 kHz) with piezoelectric ceramic (six-bolt-clamped Langevin type) transducers with the diameter of 15 mm, torsional vibrating system (20 kHz) with the diameter of 40 mm and a welding frame. Major reduction in the total needed velocity amplitude of the two-vibration-system welding machine is observed, and the welding features are enhanced notably compared to those of traditional one vibration system for plastic sheet joining.

Welding of thermoplastics is broadly carried out by using ultrasonic plastic welding machine in different industrial sectors. Welding features are enhanced effectively using high-frequency machine and two-vibration-system welding technique. One-third of velocity amplitude is used to join plastic sheets with high-frequency system

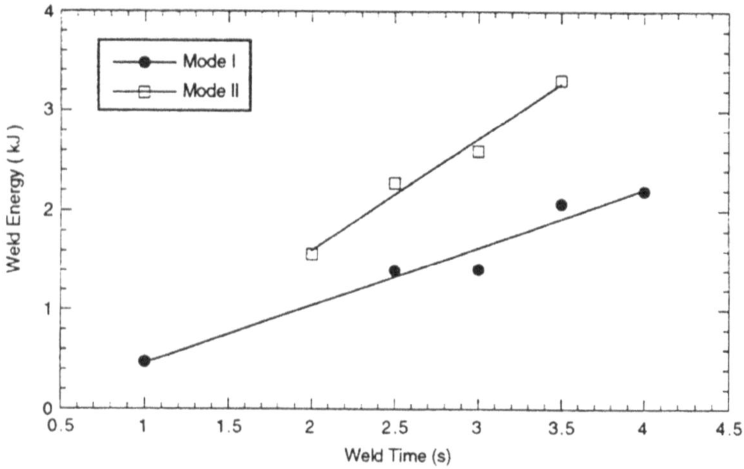

Fig. 3.10 Relationship between weld energy and weld time for mode-1 and mode-2 specimens [26]

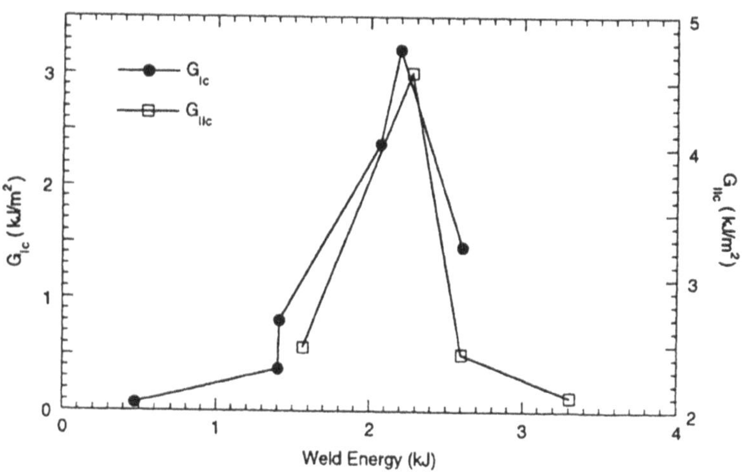

Fig. 3.11 Difference of critical strain energy release rate for mode-1 and mode-2 as a function of weld energy input [26]

(90 kHz) when compared to low-frequency system (27 kHz). Still, it is complex to join bigger samples successfully due to high-frequency vibrational stress that is unevenly distributed because of the stress relaxation impact by the small vibration displacement. Three categories of two-vibration-system ultrasonic welding machines (90 and 27 kHz longitudinal vibration systems, elliptical vibration loci of 27 kHz complex vibration systems, and 90 kHz longitudinal and 20 kHz torsional vibration systems) are used to enhance the direct welding features. Pressure sensitive films (pre-scale) are utilized to examine the transmission conditions of the vibration stresses by insert-

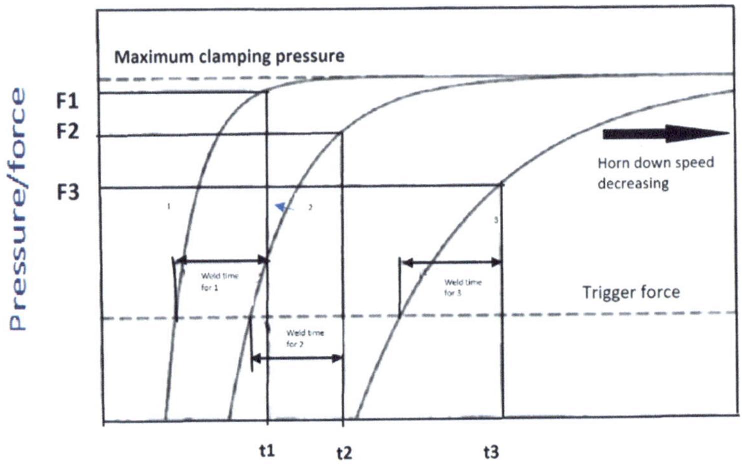

Fig. 3.12 Clamping pressure built-up diagram [29]

ing the film among the plastic sheets. It is observed that a high-frequency vibrational stress with tiny displacement amplitude may be actuated evenly in the welding samples by mixing a low-frequency vibration with large displacement amplitude. Plastic sheets are welded effectively and successfully for the welding systems [28].

Matsuoka [29] studied the comparison of the experimental results of similar and dissimilar thermoplastics weldability using ultrasonic welding and earlier method of heated tool technique. Weldability is compared with the impacts of diffusion, surface tension, solubility parameter, thermal expansion, contact pressure, and vibration transmissibility. Weld efficiency, quality of autohesion and adhesion are evaluated for the welded samples. Subsequently, comparison of the weld strength to bulk strength ratios and the ultimate extension values are carried out. The impact of contact force at the duration of welding is evaluated by varying the horn down speed in ultrasonic welding. Geometrical impacts such as the sample length and sandwich weld structure are investigated for various ultrasonic applications. Observations (Figs. 3.12, 3.13, and 3.14) are shown that weld strength is considerably reduced with respect to the threshold length of the weld samples. Besides, it appears that a sandwiched structure can be deployed to determine additional data on energy transformation and transmission for ultrasonic welding.

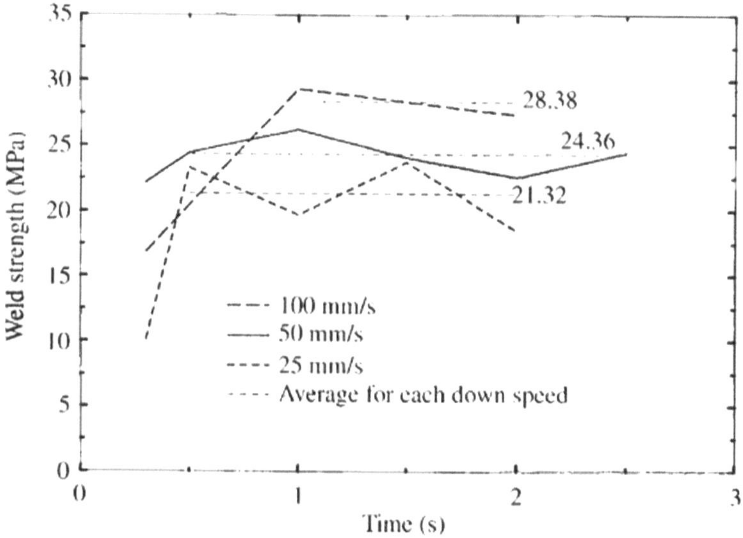

Fig. 3.13 Influence of the horn down speed on the weld strength of HS 1000 [29]

Fig. 3.14 Effect of the specimen length on the weld strength of HS 1000 [29]

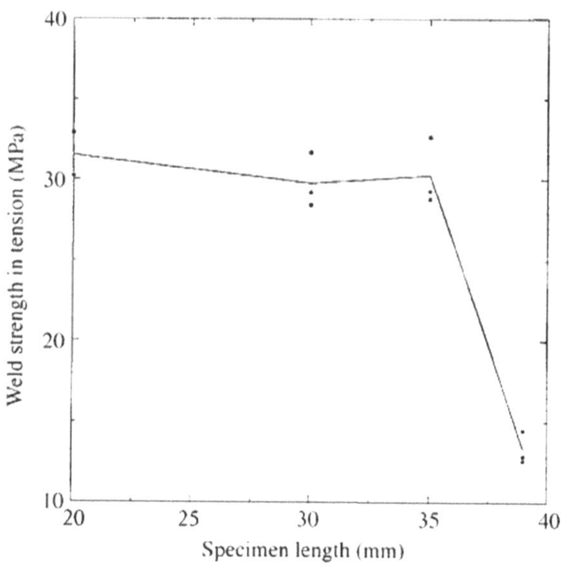

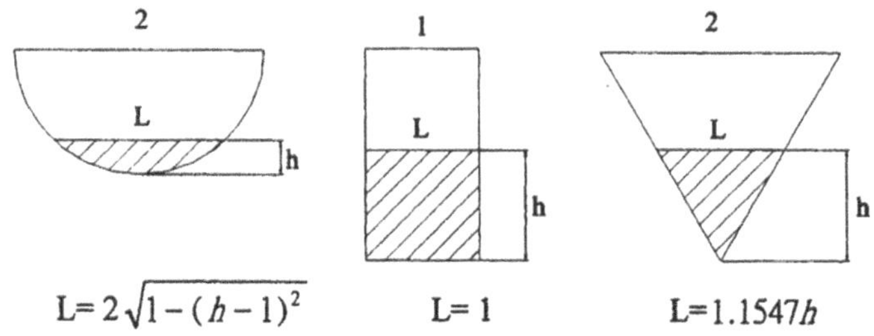

Fig. 3.15 Relationship between joint width and melted height for semicircular, rectangular and triangular energy directors, L is the joint width (mm) and h is the melted height (mm) [29]

3.2.3 Progresses in the Ultrasonic Welding During the Years of 2000s

Harras et al. [30] studied the enhancement of welding quality and speed by using energy director in ultrasonic welding. Various types of energy directors (triangular, rectangular, and semicircular) are attempted in their work. Far-field welded specimens of amorphous (ABS) and semicrystalline (PE) thermoplastics are experimented for energy director study. Weld time is the significant parameter of ultrasonic welding for all three categories of energy director which is observed. Weld pressure provides various impacts for various plastics. Welding efficiency is reduced while increasing the weld pressure for ABS. Welding efficiency is enhanced for PE while increasing the weld pressure up to four bars. Significant impact on the welding efficiency is observed with the shape of the energy director (Fig. 3.15). The results of the comparison for the shape of energy director provide higher welding efficiency on semicircular shape and lower welding efficiency on triangular shape using similar welding conditions (Fig. 3.16). Welding energy absorbed in the energy director for ABS with 48.5% and PE with 21.1.% is observed during welding process through the temperature measurements at the triangular energy director. Variations in the viscosity and elasticity among amorphous (ABS) and semicrystalline (PE) polymers are the cause for the variations in the rates of energy absorption.

Higher and fundamental resonance frequency vibrations in ultrasonic plastic welding are investigated simultaneously. Welding features are enhanced because of the larger vibration loss of polymer materials with the use of higher frequency. The higher resonance frequencies with the vibration of 95 kHz with maximum velocity of 1.5 m/s are used. The fundamental resonance frequency is used the maximum velocity of 4.5 m/s with vibration of 26 kHz. The combination of fundamental and higher resonance frequencies is used to assess the welding features of PE sheets (1 mm thickness). Increasing the higher frequency modes provides enlargement of welded area. Improvement in the welding characteristics of ultrasonic plastic welding is seen considerably while driving higher resonance frequencies [31].

Specific characteristics are replicated into large plastic plates through hot embossing. Microelectromechanical system (MEMS) and Lithographie, Galvanoformung,

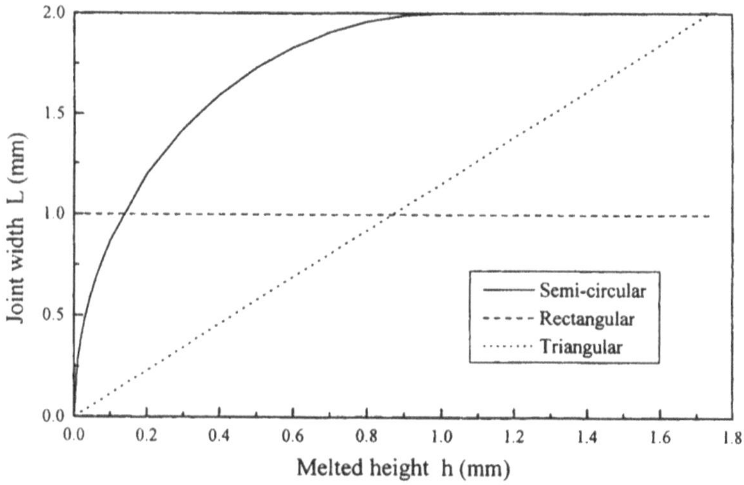

Fig. 3.16 Difference of joint width with melted height (collapse) [29]

Abformung (LIGA) technologies are used to manufacture the microstructures in master molds which are popular approaches. However, there are still few unsolved issues that startle the overall achievement of this technology. Some of the issues are long cycle times due to hot oil or conventional electric heating. This work presents the ultrasonic vibration used as heat generator for hot embossing. This investigation reports the reproduction capacity of ultrasonic-heating embossing of semicrystalline and amorphous plastic plates, the impacts of different ultrasonic vibration parameters on the contour of reflected structure, and the respective importance of all these parameters on molded component quality. Besides, the measurement is carried out for the temperature profiles at various depths of the ultrasonically vibrated-embossed plates. Temperature differences for traditional hot embossing and ultrasonic vibration hot embossing are shown in Fig. 3.17. Replicating the mold structure onto plastic plates is carried out precisely through the ultrasonic vibrated hot embossing is observed from the experimental results. Advantages of hot embossing are clearly visible from the experiments such as enhanced product quality and shorter cycle time to produce the parts [32].

Assessment of weld quality immediately after production of the joint using ultrasonic welding and monitoring the process in real time is investigated in this paper. Input electrical impedance throughout welding period is a significant factor to assess through the parameters such as input voltage and current of a welding machine in this method. Since no modifications of the dynamic properties of the welding machine and fixtures throughout the process are observed, the identified waveforms of the real and imaginary element of the impedance in a straight line replicate the thermo-mechanical performance of the plastics at the joining interface. Proper assessment of joint quality is carried out to recognize the patterns of these waveforms. Experimental investigation is performed, and subsequently, nondestructive testing process is done

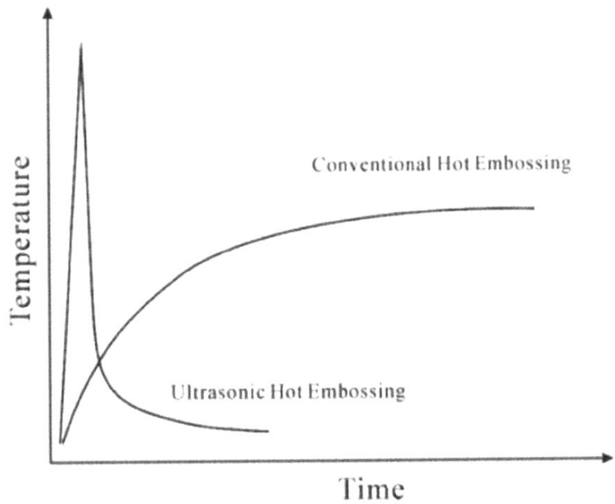

Fig. 3.17 Temperature differences for traditional hot embossing and ultrasonic vibration hot embossing [32]

for evaluating the quality. The outcome reveals that the input electrical impedance is an efficient identity and that the produced system assesses the joint quality of welds effectively [33].

Ultrasonic welding played a significant role in joining of micro-polymer parts presented in this paper. It appears to be particularly suited for the production of single material, chemically inert plastic micro-fluidic parts and systems for usages in the pharmaceutics sector, life sciences, and biotechnology as well as for high-temperature plastic micro-devices. The behavior of this 'micro-ultrasonic joining' method, alone utilizing a pneumatic standard ultrasonic machine, is exhibited by the instance of sealing and covering micro-channels and assembling a piezo-driven micro-pump [34].

Tsujino et al. [35] reported the development of the finite element model for ultrasonically welded continuous carbon fiber (CF)-reinforced polyether ether ketone (PEEK) sheets using viscoelastic dissipation theory in this work. The influence of the energy director geometry such as size and apex angle during heating process in the ultrasonic welding is recorded. The outcome of the simulation reveals that the major important influence is provided by the apex angle of the energy director for temperature rising rate than the energy director size (Fig. 3.18). Temperature filed profile of the energy director is mostly influenced by the apex angle. Butterfly shape of the energy director at the highest temperature region and deviation from the lower sheet is observed while the apex angle with 30° and 60°. Lower sheet and energy director with apex angle of 90° (the 'wings' of the highest temperature region) are almost merged together. Closed ellipse area or wrapped in the energy director tip is observed for the highest temperature region of energy director with apex angle of 120°. On completion of heating during the ultrasonic joining of samples, it may be observed that the cross sectional area of the energy director reduces prominently as compared to its original size. The recorded cross sectional area of the energy director after it gets subjected to complete heating is 0.25 mm^2. The heated region

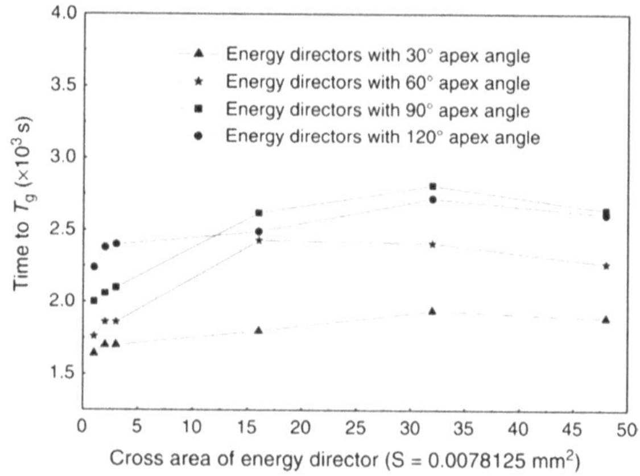

Fig. 3.18 Time to Tg of the energy directors with various energy director sizes and apex angles [35]

on the energy director and lower sheet area are constant while an increase in the cross-sectional area (more than 0.25 mm^2). As a result, most appropriate apex angle for energy director is 90°, and most suitable cross-sectional area is about 0.25 mm^2 for the ultrasonic welding of CF/PEEK sheets.

The field of structural and non-structural applications mostly demands the use of engineering plastics rapidly. Hence, joining of plastics is essential to manufacture the plastic components. From the various methods of thermoplastics joining, ultrasonic welding is mostly preferred process. Different types of thermoplastics such as amorphous and semicrystalline are needed for different method to join using ultrasonic welding. Heating region is restricted only on the interface, hence, weld quality majorly based on the temperature at the interface. Therefore, temperature distribution on the interface is most significant to investigate and predict the weld quality. Heat is established at the interface due to viscoelastic heating mechanism in the ultrasonic welding of thermoplastics. Viscoelastic heating is based on the parameters such as square of amplitude, applied frequency, and loss modulus. In this research, ANSYS tool is used to model and simulate the results of the temperature distribution for different joint designs of thermoplastics components. Validation of the results is performed among the model and measurement of temperature during welding [36].

Ultrasonic welding of thermoplastics is discussed with the calculation of energy, weld quality, properties of welded components and governing parameters. In particular, parameters such as weld time, amplitude, frequency of ultrasonic welding, and thickness of materials are discussed. From the observation of the ultrasonic welding, it is found that the weld energy required for joining of plastics and weld time is majorly influenced the thermoplastic materials in this study [37].

MEMS packaging is significant application in the polymer MEMS field where joining of MEMS is performed with ultrasonic welding machine. Various benefits such as no filler materials or additional heating source required, short cycle time, minimum damage to specimens, very little heat affected zone are the important reasons for choosing the ultrasonic welding of thermoplastics. Ultrasonic welding

(USW) used in the field-off polymer MEMS packaging produces an afresh technology named ultrasonic bonding (USB). Application range of USW is broadened for the polymer MEMS devices from the range of macro-device–micro-device for thermoplastics. The polymer MEMS devices used to join with the development of USW and USB are discussed briefly in this work. Afterward, several PMMA microfluidic chips with ultrasonically joined parts using MEMS production technology are reported. USB experiment is performed with ultrasonic plastic welding machine based on these chips. The joint accomplished poses high strength. Though there are several issues in joining of polymer, MEMS is preferred due to miniaturization of structure of polymer components [38].

The practicality of ultrasonic bonding is demonstrated for MEMS packaging using the solid phase vibration and joining of the two parts at lower temperature quickly. Vertical ultrasonic bonding and lateral ultrasonic bonding are the two different approaches with three pairs of materials such as In–Au, Al–Al, and plastics to plastics in this study. The process uses only mechanical vibration as energy to join low-temperature welding among similar and dissimilar materials without cleaning the joining surfaces. The ultrasonic bonding of MEMS packages is completed within second using power of tens of watts at room temperature with area of a few mm^2. Gross leakage tests are carried out successfully for the packaged MEMS where bonded chips are immersed into the liquid. And there is no leakage found in the bonded chips [39].

3.2.4 Progresses in the Ultrasonic Welding During the Years of 2010s

The paper discusses the outcome of theoretical studies of creation of continuous weld in contact region of planar surface of the radiator with rotating plane of the pressure roller. The suggestions are made from this study such as selection of pressure roller parameters, optimal amplitude of ultrasonic vibrations, and broach speed for materials with various properties and thickness. These are established on the basis of distribution analysis, replication of properties, and absorption of vibrations from ultrasonic machine in the weld interface [40].

Numerical simulation and experimental investigations for the heating mechanisms of ultrasonic welding are studied in this work. Theoretical analysis is carried out for the relationship among static relaxation and dynamic modulus of thermoplastics. Simulation is performed using the segment time–temperature equivalency equation for various temperature series and a technique for directly dividing viscoelastic heat from strain energy. Validation of simulation results is performed by using temperature measurement tests with poly(methyl methacrylate) samples. Initially, interfacial friction initiating the welding process of thermoplastics is observed from the outcome of simulation and experiment (Fig. 3.19). Viscoelastic heating plays a dominant role

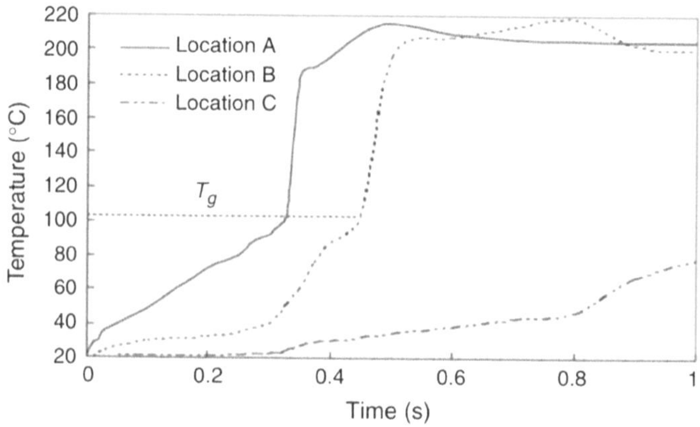

Fig. 3.19 Temperature curves of specimen with interface obtained from experiment [41]

after reaching the temperature of glass transition (Tg) of the material. And sufficient heat is supplied through the viscoelastic heat during welding [41].

Welding of continuous fiber-reinforced thermoplastic composites is a challenging task for the engineers which are carried out most successfully through ultrasonic welding method. Ultrasonic waves applied on the surfaces yield intermolecular friction within the bulk and produce the heat needed for joining at the interface of the bonding members through energy directors (EDs). Energy directors comprise protrusions with different shapes (triangular, rectangular, and semicircular) on the material surfaces and play a significant role in the welding process and weld quality. This paper reveals the outcome of an investigation on the impacts of various EDs of carbon fiber/polyetherimide advanced thermoplastic composites using near-field ultrasonic welding. Composite sheets are fabricated and molded with triangular EDs using hot platen press. The effects of the orientation of EDs (various shape) relating to the load direction are evaluated using single-lap shear welded specimens. The outcome specifies that the arrangement of different EDs is more efficient in covering the overlap region, and once the resin has melted locally, it produces a minimum fiber disruption at the joint interface [42].

Kim et al. [43] proposed and investigated the novel welding method to join the heterogeneous materials of ductile polycarbonate (PC) and poly(methyl methacrylate) (PMMA) by achieving the strength and various welding conditions. Functional gradient interposed sheets (IPS) fabricated from three layers is valid for bond along with its exceptional mechanical properties. Air bubbles and IPS cracking among the interfaces are produced due to excessive weld time. Enhancement of weld strength as well as air bubble removal is achieved through higher welding stress which also produces enlargement of weld area. Hence, improvement of welding strength is carried out by enlarging welding region and lessening the air bubbles. The study reveals that the efficient welding conditions are the higher welding stress and shorter duration of welding. From this investigation, the heterogeneous materials are welded effectively using this novel welding technique.

Fabrication of thermoplastic composite parts involves ultrasonic plastic welding. Weld interface of the plastic parts is welded by high-frequency vibration that produces self-heating and localized melting. The principle characteristic of this method is the existence of phenomena that take place on two varied time scales like the flow of molten plastics about 1 s and the vibration about 10–5 s. Time homogenization method is applied to simulate these phenomena precisely without using very fine time discretization over the entire course. Initially, Maxwell viscoelastic constitutive law is used to formulate the thermomechanical problem, and consequently, asymptotic expansion is utilized to homogenize. This directs to three tied issues like macro-chronological mechanical, micro-chronological mechanical, and macro-chronological thermal problems. This tied formulation is really simpler for the reason that macro-chronological problems are not based on the micro-timescale and are related to fast differences. Finally, proposal of a uniform simple test case is carried out for comparing the homogenized solution to a direct computation. It reveals that the technique offers excellent outcome, provided that the vibration is quicker than the entire process duration. Besides, the time of the process is lessened by 1000 times than earlier processes [44].

Ultrasonically welded acrylonitrile butadiene styrene (ABS) and poly(methyl methacrylate) (PMMA) parts are investigated for optimizing the process parameters (weld pressure, vibration amplitude, and welding time) using the model created through artificial intelligence (AI) techniques in this paper. ABS and PMMA specimens are welded using ultrasonic spot welding experimentally. Artificial intelligence tools such as adaptive fuzzy inference systems, artificial neural network (ANN), and hybrid systems are used to create model and optimization for the conducted experiments. Predictions of ANN are performed better than other AI techniques. Process parameters are precisely predicted through the feed-forward back-propagation network by means of uniform transfer functions (4/2 neurons in the 1/2 hidden layers). Predicted values are provided into the particle swarm optimization (PSO) and genetic algorithm (GA) for optimizing the process parameters. Similar outcome of GA and PSO is seen after optimization and the computed strength enhanced by 10% as compared with non-optimized welded parts. Agreeable results are provided by the optimization which is confirmed by the validation of the computed and experimental outcome [45].

Glass fiber-reinforced polypropylene composites welded using ultrasonic welding machine is investigated in this paper. Strength of welded parts is evaluated through the influence of process parameters such as pressure, time, and amount of reinforcement. Response surface methodology (Box–Behnken test) is employed to lessen the number of experimental trials and cost with the influence of varying parameters on strength. Optimization is performed through Box–Behnken test by pondering the parameters with four factors at three levels to achieve the utmost level of strength. The outcome of this investigation revealed that amplitude of 32 μ (more than mean value), 1.5 times higher air pressure, maximum failure force of 2.30 KN, utilization of glass fiber (10%), and welding duration of 0.4 s yields optimized weld [46].

Ultrasonic welding is espoused to join the thermoplastic composites (material with energy director) due to their advantages such as faster welding process, no filler materials, and good quality of joints. This work reports one additional benefit, the capability to link weld strength with the welding process data, specifically power required and displacement of the horn in ultrasonic thermoplastics welding with flat energy director. This connection with the combination of displacement-controlled welding permits optimization of the welding parameters which constantly yield high strength joints [47].

Proposal of a model for the oscillating deformation (mechanics), friction dissipation and viscoelastic heat creation, and intimate conduct and healing (degree of adhesion) is made for the first stage of transient heating in this study. Multi-physical finite element analysis is espoused to provide numerical solution. Earlier, experimental values for delivered power correlated with predicted power dissipation are observed and achieved with global efficiency of 13% in the apparatus. Adhesion is initiated on the edges of the sample initially and gradually progresses across the entire lap interface region which is confirmed through prediction studies. Physical mechanisms taking place and duration that required for the entire welding process are studied using numerical model. Literature reports that the interfacial friction provides the first heating mechanism for joining parts. After reaching maximum temperatures, bulk viscoelastic dissipation plays dominant role in joining of parts. The dissipation of the power is rapidly amplified after reaching glass transition temperature in between the entire interface [48].

Welding of thermoplastics is dominantly produced through the energy directors (responsible for local heat generation) which are molded on the surfaces as protrusions. This investigation assesses a substitute energy director (flat energy director) for ultrasonic thermoplastics welding. Flat energy director provides the adequate welding strength which is assessed through the displacement of horn, dissipated power, weld time, and welding energy for the ultrasonically welded thermoplastic composites (Fig. 3.20), [49]. Similar welding strength of thermoplastic composite specimen with flat energy director and the conventional energy director specimens is observed through the experiments [50, 51].

3.3 Significance of Ultrasonic Welding

Ultrasonic plastic welding produces permanent joint between interfaces of plastic components. Permanent bond is generated among the plastic components by frictional heat through mechanical vibrations, and pressure which causes local melting of the components. The metal parts embedded into plastic parts are the significant merit of ultrasonic welding. Replacement of metal is emerging in the automobile industry where fusing of plastic into metal is a significant problem. Ultrasonically welded plastic parts are employed in the automobile sectors from steering wheels to door panels (Fig. 3.21). Merits of the ultrasonic thermoplastic welding are automation, low cycle times, and flexibility. Besides, ultrasonic thermoplastic welding does

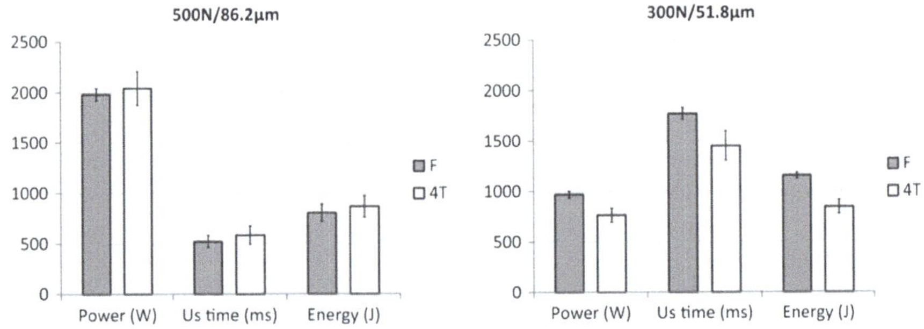

Fig. 3.20 Maximum power, vibration time, and welding energy for two different sets of welding parameters with two types of energy director [50]

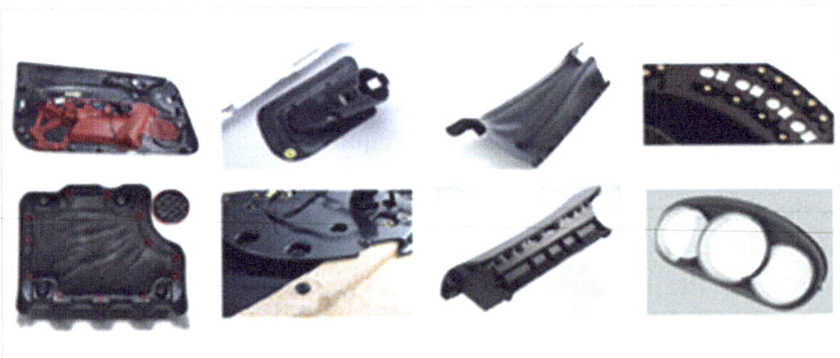

Fig. 3.21 Ultrasonically welded automobile parts

not smash up surface finish of the product due to their high-frequency vibrations to avoid scratches from being produced. Defects can influence the car manufacturer's bottom line which is a significant concern. Dissimilar combination of materials is welded using ultrasonic joining technique such as semiconductor materials, high conductivity, and temperature materials (gold, copper, aluminum, silver, etc.).

Medical industry is benefitted from the ultrasonic thermoplastic welding due to their contamination-free products. Ultrasonic thermoplastic welding provides the products such as blood filters, anesthesia filters, dialysis tubes, IV catheters, and face masks (Fig. 3.22). One more significant usage in the medical domain is the fabrication of such textile products like surgery garments, hospital gowns, and transdermal patches. These products are sealed and sewn by using ultrasonic welding to prevent and lessen the possibility of infection.

Packaging sector is also benefitted from the ultrasonic plastic welding regularly. Everyday products (from butane lighters to food containers) are made or fabricated using ultrasonic thermoplastic welding (Fig. 3.23). Packaging of hazardous explosives such as firecrackers or hazardous chemicals is produced from ultrasonic plastic welding.

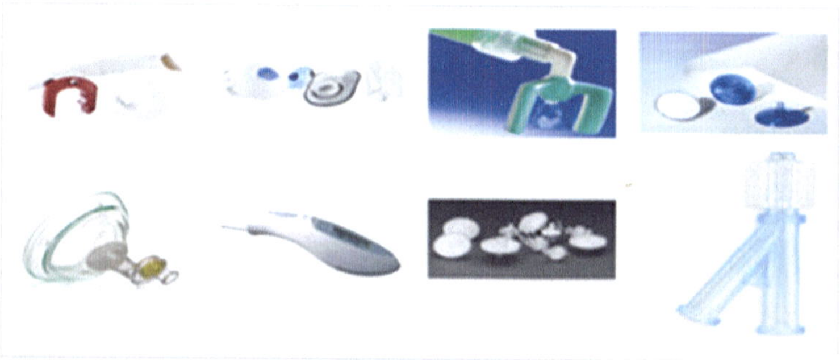

Fig. 3.22 Ultrasonically welded medical equipments

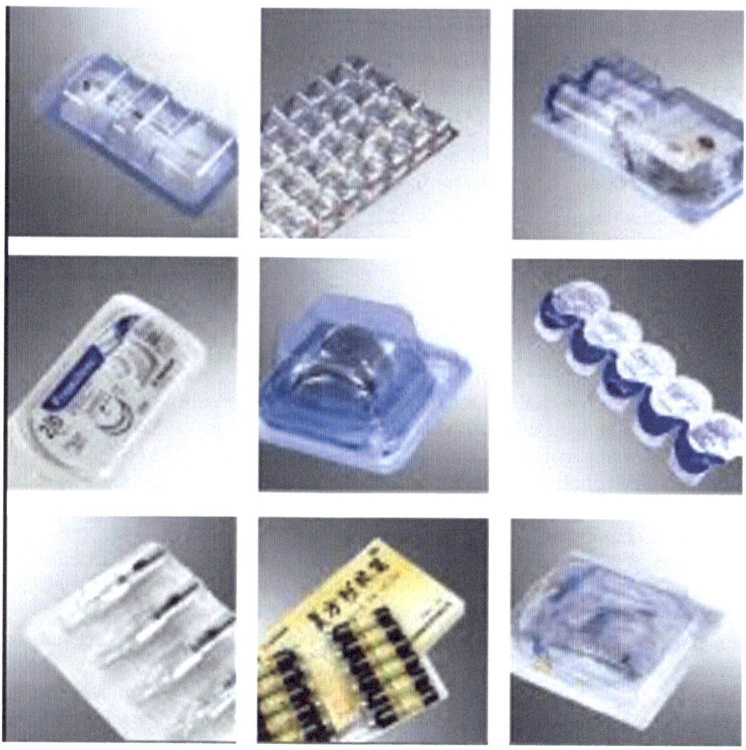

Fig. 3.23 Packed products using ultrasonic welding

Airtight seals of the food products in the food industry are made using ultrasonic plastic welding due to their clean and fast process. Major instances of products sealed using ultrasonic plastic welding machine are juice and milk containers (Fig. 3.24).

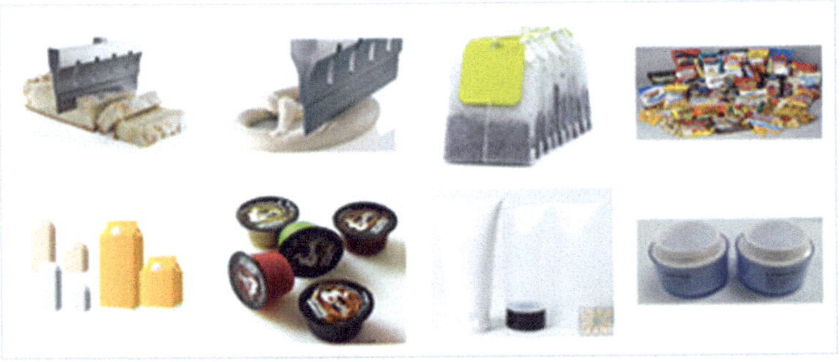

Fig. 3.24 Ultrasonically welded food products

Power consumption of the ultrasonic plastic welding is lesser than the resistance spot welding. Strength of the joint is high and does not require higher cleanliness of the surface of the plastics.

Ultrasonic welding is broadly employed to produce the microelectronic equipments by making interconnection between the integrated circuit elements and is also utilized in the electrical packaging industries.

References

1. Truckenmüller, Roman, Ralf Ahrens, Yue Cheng, Günther Fischer, and Volker Saile. 2006. An ultrasonic welding based process for building up a new class of inert fluidic microsensors and-actuators from polymers. *Sensors and Actuators, A: Physical* 132 (1): 385–392.
2. Fernandez Villegas, I., B. Valle Grande, H.E.N. Bersee, and R. Benedictus. 2015. A comparative evaluation between flat and traditional energy directors for ultrasonic welding of CF/PPS thermoplastic composites. *Composite Interfaces* 22 (8): 717–729.
3. Stoehr, Neda, Benjamin Baudrit, Edmund Haberstroh, Michael Nase, Peter Heidemeyer, and Martin Bastian. 2015. Ultrasonic welding of plasticized PLA films. *Journal of Applied Polymer Science* 132 (4).
4. Sackmann, J., K. Burlage, C. Gerhardy, B. Memering, S. Liao, and W.K. Schomburg. 2015. Review on ultrasonic fabrication of polymer micro devices. *Ultrasonics* 56: 189–200.
5. Rizzolo, Robert H., and Daniel F. Walczyk. 2015. Ultrasonic consolidation of thermoplastic composite prepreg for automated fiber placement. *Journal of Thermoplastic Composite Materials*. https://doi.org/10.1177/0892705714565705.
6. Rani, M. Roopa, K. Prakasan, and R. Rudramoorthy. 2015. Studies on thermo-elastic heating of horns used in ultrasonic plastic welding. *Ultrasonics* 55: 123–132.
7. Patrikios, Michael, Robert S. Soloff, and Sigfredo Vargas. 2015. *System and method for mounting ultrasonic tools*. U.S. Patent 8,950,458, issued February 10.
8. Kistrup, Kasper, Carl Esben Poulsen, Mikkel Fougt Hansen, and Anders Wolff. 2015. Ultrasonic welding for fast bonding of self-aligned structures in lab-on-a-chip systems. *Lab on a Chip* 15 (9): 1998–2001.
9. Deng, Shiqiang, Luke Djukic, Rowan Paton, and Lin Ye. 2015. Thermoplastic–epoxy interactions and their potential applications in joining composite structures—A review. *Composites Part A Applied Science and Manufacturing* 68: 121–132.

10. Levy, Arthur, Steven Le Corre, and Irene Fernandez Villegas. 2014. Modeling of the heating phenomena in ultrasonic welding of thermoplastic composites with flat energy directors. *Journal of Materials Processing Technology* 214 (7): 1361–1371.
11. Villegas, Irene Fernandez. 2014. Strength development versus process data in ultrasonic welding of thermoplastic composites with flat energy directors and its application to the definition of optimum processing parameters. *Composites Part A Applied Science and Manufacturing* 65: 27–37.
12. Rani, M. Roopa, K. Prakasan, and R. Rudramoorthy. 2014. Design and simulation of a large tubular horn for ultrasonic plastic welding. *International Journal of Design Engineering* 5 (4): 344–357.
13. Benatar, Avraham, and Timothy G. Gutowski. 1989. Ultrasonic welding of PEEK graphite APC-2 composites. *Polymer Engineering and Science* 29 (23): 1705–1721.
14. Nonhof, C.J., and G.A. Luiten. 1996. Estimates for process conditions during the ultrasonic welding of thermoplastics. *Polymer Engineering and Science* 36 (9): 1177–1183.
15. Lee, Woo Il, and George S. Springer. 1987. A model of the manufacturing process of thermoplastic matrix composites. *Journal of composite materials* 21 (11): 1017–1055.
16. Butler, Christine A., Roy L. Mccullough, Ranga Pitchumani, and John W. Gillespie. 1998. An analysis of mechanisms governing fusion bonding of thermoplastic composites. *Journal of Thermoplastic Composite Materials* 11 (4): 338–363.
17. Levy, Arthur, Steven Le Corre, and Arnaud Poitou. 2014. Ultrasonic welding of thermoplastic composites: a numerical analysis at the mesoscopic scale relating processing parameters, flow of polymer and quality of adhesion. *International Journal of Material Forming* 7 (1): 39–51.
18. Carboni, Michele. 2014. Failure analysis of two aluminium alloy sonotrodes for ultrasonic plastic welding. *International Journal of Fatigue* 60: 110–120.
19. Suresh, K. S., M. Roopa Rani, K. Prakasan, and R. Rudramoorthy. 2007. Modeling of temperature distribution in ultrasonic welding of thermoplastics for various joint designs. *Journal of Materials Processing Technology* 186 (1): 138–146.
20. Tsujino, Jiromaru, Misugi Hongoh, Ryoko Tanaka, Rie Onoguchi, and Tetsugi Ueoka. 2002. Ultrasonic plastic welding using fundamental and higher resonance frequencies. *Ultrasonics* 40 (1): 375–378.
21. Frankel, E.J., and K.K. Wang. 1980. Energy transfer and bond strength in ultrasonic welding of thermoplastics. *Polymer Engineering and Science* 20 (6): 396–401.
22. Nikoi, R., M.M. Sheikhi, and N.B.M. Arab. 2014. Experimental analysis of effects of ultrasonic welding on weld strength of polypropylene composite samples. *International Journal of Engineering-Transactions C: Aspects* 28 (3): 447–453.
23. Levy, A., S. Le Corre, and I.F. Villegas. 2014. Modeling of the heating phenomena in ultrasonic welding of thermoplastic composites with flat energy directors. *Journal of Materials Processing Technology* 214 (7): 1361–1371.
24. Tolunay, M.N., P.R. Dawson, and K.K. Wang. 1983. Heating and bonding mechanisms in ultrasonic welding of thermoplastics. *Polymer Engineering and Science* 23 (13): 726–733.
25. Potente, H. 1984. Ultrasonic welding—Principles and theory. *Materials and Design* 5 (5): 228–234.
26. Benatar, A., and Z. Cheng. 1989. Ultrasonic welding of thermoplastics in the far-field. *Polymer Engineering and Science* 29 (23): 1699–1704.
27. Benatar, A., and T.G. Gutowski. 1989. Ultrasonic welding of PEEK graphite APC-2 composites. *Polymer Engineering and Science* 29 (23): 1705–1721.
28. Benatar, A., R.V. Eswaran, and S.K. Nayar. 1989. Ultrasonic welding of thermoplastics in the near-field. *Polymer Engineering and Science* 29 (23): 1689–1698.
29. Matsuoka, S.I. 1995. Ultrasonic welding and characteristics of glass-fiber reinforced plastic: comparison between the paper-making method and the impregnation method. *Journal of Materials Processing Technology* 55 (3–4): 427–431.
30. Harras, B., K.C. Cole, and T. Vu-Khanh. 1996. Optimization of the ultrasonic welding of PEEK-carbon composites. *Journal of Reinforced Plastics and Composites* 15 (2): 174–182.

31. Tsujino, J., T. Tamura, T. Uchida, and T. Ueoka. 1996. Characteristics of two-vibration-system ultrasonic plastic welding with 90 kHz and 20 kHz vibration systems at right angles. *Japanese Journal of Applied Physics* 35 (11R): 5884.
32. Tsujino, J., T. Uchida, K. Ohkusa, T. Adachi, and T. Ueoka. 1998. Transmission conditions of vibration stresses to welding specimens of ultrasonic plastic welding using various two-vibration-system equipments. *Japanese Journal of Applied Physics* 37 (5S): 3001.
33. Sancaktar, E. 1999. Polymer adhesion by ultrasonic welding. *Journal of Adhesion Science and Technology* 13 (2): 179–201.
34. Chuah, Y.K., L.H. Chien, B.C. Chang, and S.J. Liu. 2000. Effects of the shape of the energy director on far-field ultrasonic welding of thermoplastics. *Polymer Engineering and Science* 40 (1): 157–167.
35. Tsujino, J., M. Hongoh, R. Tanaka, R. Onoguchi, and T. Ueoka. 2002. Ultrasonic plastic welding using fundamental and higher resonance frequencies. *Ultrasonics* 40 (1–8): 375–378.
36. Liu, S.J., and Y.T. Dung. 2005. Hot embossing precise structure onto plastic plates by ultrasonic vibration. *Polymer Engineering and Science* 45 (7): 915–925.
37. Ling, S.F., J. Luan, X. Li, and W.L.Y. Ang. 2006. Input electrical impedance as signature for nondestructive evaluation of weld quality during ultrasonic welding of plastics. *NDT and E International* 39 (1): 13–18.
38. Truckenmüller, R., Y. Cheng, R. Ahrens, H. Bahrs, G. Fischer, and J. Lehmann. 2006. Micro ultrasonic welding: Joining of chemically inert polymer microparts for single material fluidic components and systems. *Microsystem Technologies* 12 (10–11): 1027–1029.
39. Wang, X., J. Yan, R. Li, and S. Yang. 2006. FEM investigation of the temperature field of energy director during ultrasonic welding of PEEK composites. *Journal of Thermoplastic Composite Materials* 19 (5): 593–607.
40. Suresh, K.S., M.R. Rani, K. Prakasan, and R. Rudramoorthy. 2007. Modeling of temperature distribution in ultrasonic welding of thermoplastics for various joint designs. *Journal of Materials Processing Technology* 186 (1–3): 138–146.
41. Khmelev, V.N., Slivin, A.N. and Abramov, A.D. 2007. Model of process and calculation of energy for a heat generation of a welded joint at ultrasonic welding polymeric thermoplastic materials. In *8th Siberian Russian workshop and tutorial on electron devices and materials, 2007. EDM'07*. 316–322. IEEE.
42. Zongbo, Z., L. Yi, W. Xiaodong, and W. Liding. 2008. Advances in ultrasonic welding of plastics and its usage in polymer MEMS bonding [J]. *Welding and Joining* 8: 007.
43. Kim, J., B. Jeong, M. Chiao, and L. Lin. 2009. Ultrasonic bonding for MEMS sealing and packaging. *IEEE Transactions on Advanced Packaging* 32 (2): 461–467.
44. Khmelev, V.N., Slivin, A.N., Lehr, A.V. and Abramov, A.D. 2010. Theoretical investigations of continuous ultrasonic seam welding of thermoplastic polymers and fabrics. In *2010 international conference and seminar on micro/nanotechnologies and electron devices (EDM)*. 341–344. IEEE.
45. Zhang, Z., X. Wang, Y. Luo, Z. Zhang, and L. Wang. 2010. Study on heating process of ultrasonic welding for thermoplastics. *Journal of Thermoplastic Composite Materials* 23 (5): 647–664.
46. Villegas, I.F., and H.E. Bersee. 2010. Ultrasonic welding of advanced thermoplastic composites: An investigation on energy-directing surfaces. *Advances in Polymer Technology* 29 (2): 112–121.
47. Qiu, J., G. Zhang, M. Asao, M. Zhang, H. Feng, and Y. Wu. 2010. Study on the novel ultrasonic weld properties of heterogeneous polymers between PC and PMMA. *International Journal of Adhesion and Adhesives* 30 (8): 729–734.
48. Levy, A., Le Corre, S., Poitou, A. and Soccard, E., 2011. Ultrasonic welding of thermoplastic composites: modeling of the process using time homogenization. *International Journal for Multiscale Computational Engineering* 9(1).
49. Villegas, I.F. 2014. Strength development versus process data in ultrasonic welding of thermoplastic composites with flat energy directors and its application to the definition of optimum processing parameters. *Composites Part A Applied Science and Manufacturing* 65: 27–37.

50. Norouzi, A., M. Hamedi, and V.R. Adineh. 2012. Strength modeling and optimizing ultrasonic welded parts of ABS-PMMA using artificial intelligence methods. *The International Journal of Advanced Manufacturing Technology* 61 (1–4): 135–147.
51. Fernandez Villegas, I., B. Valle Grande, H.E.N. Bersee, and R. Benedictus. 2015. A comparative evaluation between flat and traditional energy directors for ultrasonic welding of CF/PPS thermoplastic composites. *Composite Interfaces* 22 (8): 717–729.

Chapter 4
Testing and Evaluation of Polymer Welds—An Insight into the Common Techniques

4.1 Mechanical Terminologies for Testing Polymer Welds

Mechanical deformation of the plastics depends on the temperature, time and viscoelastic nature. The duration of the applied stress and the overall stress history are the stages dictating time dependency. Temperature dependent plastics deformation is affected by the thermal properties of the plastics which typically vary for thermoplastics and thermosets. Temperature dependency basically controls the time for plastic deformation. Semicrystalline or glassy plastics exhibits weaker viscoelastic nature at temperatures below their glass transition temperatures (Tg). It is essential to analyze the semicrystalline plastics for time-dependency based analysis owing to their nature. Increased mechanical response during time analysis is recorded due to the increased temperature either by heat provided during deformation or by the external heat fluxes. Plastics reveal significant difference in the deformation mechanism as compared to metals.

In plastics, heat is the key factor affecting plastic deformation. Differences in arrangement of the lengthy chain molecules about their centre of gravity produce the damage in the plastics because of cavitation and non-cavitation systems. In polymers, whole molecule is involved in governing the deformation process to settle about its center of gravity, while in metals, the deformation is predominantly governed by the alterations in positions and corresponding orientations of the center of gravity about its molecules. Stored internal energy being high throughout the plastic deformation process yields various impacts which are not observed in case of metals.

Plastically deformed specimen is heated in the nonexistence of external constraints which will contract towards its original length. Without reaching the melting point, plastic deformation is fully recoverable in most of the polymers. Constant length of the plastically deformed specimen during heating is observed and significant increase in stress is observed. These experiments evidently reveal the variation in the plasticity of polymers and plasticity of metals.

S. A. Vendan et al., *Confluence of Multidisciplinary Sciences for Polymer Joining*,
https://doi.org/10.1007/978-981-13-0626-6_4

Table 4.1 Mechanical properties of thermoplastics at room temperature [1]

Material	Tensile strength (MPa)	Elongation (%)	Modulus of elasticity (GPa)	Compressive strength (MPa)	Hardness (HV)
ABS (acrylonitrile-butadiene-styrene)	35–45 (5–7)	15–60	1.7–2.2 (0.25–0.32)	–	95–105 HRR
CA (cellulose acetate)	15–60 (2–9)	6–50	0.6–3.0 (0.1–0.4)	90–250 (13–36)	50–125 HRR
CN (cellulose nitrate)	50–55 (7–9)	40–45	1.3–15.0 (0.18–2)	150–240 (22–35)	95–115 HRR
PA (polyamide)	80 (12)	90	3.0 (0.43)	85 (12)	79 HRM, 118 HRR
PMMA (polymethyl methacrylate)	50–70 (7–10)	2–10	–	80–115 (12–17)	85–105 HRM
PS (polystyrene)	35–60 (5–9)	1–4	3.0–4.0 (0.4–0.6)	80–110 (12–16)	65–90 HRM
PVC (polyvinyl chloride), rigid	40–60 (6–9)	5	2.4–2.7 (0.3–0.4)	60 (9)	110–120 HRR
PVCAc (polyvinyl chloride acetate), rigid	50–60 (7–9)	–	2.0–3.0 (0.3–0.4)	70–80 (10–12)	–

Mechanical properties are listed for several plastics in the Table 4.1 [1], and this section briefly provides various techniques adopted for testing of mechanical properties. Mechanical properties in general are investigated as per ASTM and ISO standards for the plastics (Table 4.2) [2]. Readers are recommended for further explanations on characterization techniques from [2] and [3].

4.2 Tensile Tests

Tensile modulus and tensile elongation measurements are between the most significant hints of strength in a material and are the most important properties of polymers. Tensile test is to identify the maximum withstanding capacity, plastic deformation and breaking point of the material while applying the load. Stress-strain diagram (Fig. 4.1) is used to examine the relative stiffness of a material which is tensile modulus. Various categories of plastic materials are often evaluated on the basis of tensile modulus, tensile strength and elongation data. Some class of

Table 4.2 ASTM and ISO mechanical test standards for plastics [2]

ASTM standard	ISO standard	Key focus
D 618	291	Methods of specimen conditioning
D 955	294-4	Measuring shrinkage from mold dimensions of molded thermoplastics
D 3419	10724	In-line screw-injection molding of test specimens from thermosetting compounds
D 3641	294-1, 2, 3	Injection molding test specimens of thermoplastic molding and extrusion materials
D 4703	293	Compression molding thermoplastic materials into test specimens, plaques, or sheets
D 524	95	Compression molding test specimens of thermosetting molding compounds
D 6289	2577	Measuring shrinkage from mold dimensions of molded thermosetting plastics
D 256	180	Determining the pendulum impact resistance of notched specimens of plastics
D 638	527-1, 2	Tensile properties of plastics
D 695	604	Compressive properties of rigid plastics
D 785	2039-2	Rockwell hardness of plastics and electrical insulating materials
D 790	178	Flexural properties of unreinforced and reinforced plastics and insulating materials
D 882	527-3	Tensile properties of thin plastic sheeting
D 1043	458-1	Stiffness properties of plastics as a function of temperature by means of a torsion test
D 1044	9352	Resistance of transparent plastics to surface abrasion
D 1708	6239	Tensile properties of plastics by use of microtensile specimens
D 1822	8256	Tensile-impact energy to break plastics and electrical insulating materials
D 1894	6601	Static and kinetic coefficients of friction of plastic film and sheeting
D 1922	6383-2	Propagation tear resistance of plastic film and thin sheeting by pendulum method
D 1938	6383-1	Tear propagation resistance of plastic film and thin sheeting by a single tear method
D 2990	899-1, 2	Tensile, compressive, and flexural creep and creep-rupture of plastics
D 3763	6603-2	High-speed puncture properties of plastics using load and displacement sensors
D 4065	6721-1	Determining and reporting dynamic mechanical properties of plastics
D 4092	6721	Dynamic mechanical measurements on plastics

(continued)

Table 4.2 (continued)

ASTM standard	ISO standard	Key focus
D 4440	6721-10	Rheological measurement of polymer melts using dynamic mechanical procedures
D 5023	6721-3	Measuring the dynamic mechanical properties of plastics using three-point bending
D 5026	6721-5	Measuring the dynamic mechanical properties of plastics in tension
D 5045	572	Plane-strain fracture toughness and strain energy release rate of plastic materials
D 5083	3268	Tensile properties of reinforced thermosetting plastics using straight-sided specimens
D 5279	6721	Measuring the dynamic mechanical properties of plastics in torsion

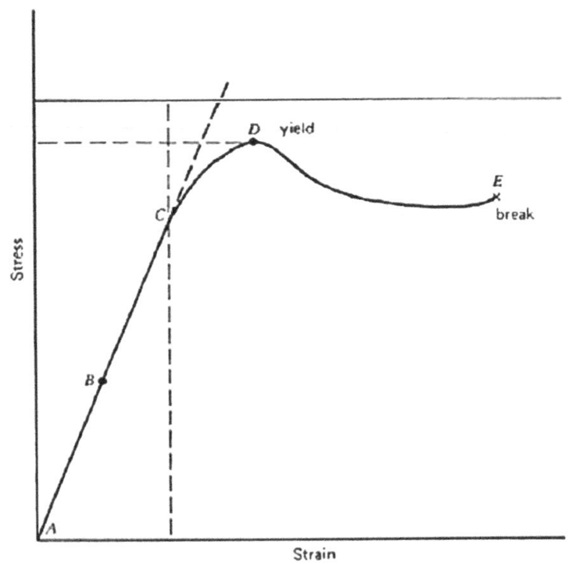

Fig. 4.1 Typical stress-strain curve [4]

plastics are susceptible to the rate of straining and other external physical/chemical factors. Hence, the experimental values determined by this technique cannot be considered valid for usages involving load-time scales or atmospheres broadly varied from this technique. Tensile properties values are more significant while choosing a category of plastic based on applications from a large group of plastic materials and such values are partially used in actual design of the product. This is due to the test does not take into consideration the time-dependent characteristic of plastic materials. Test procedures vary depending upon the test specimens and according ASTM standard and procedures adopted also varies. Important parameters influencing tensile results are specimen geometry, rate of straining and temperature [4].

Fig. 4.2 Tensile testing machine

Constant rate crosshead movable tensile testing machine is employed. Tensile machine comprises of two fixtures to hold the work piece, where one end is fixed and other end is movable. The alignment problem of the specimen in the tensile testing machine is eliminated through the self-aligning grips. Controlled movement is carried out using a controlled-velocity drive mechanism. High precision speed control is achieved through several machines available in the market such as closed-loop servo-controlled drive mechanism. A load-specifying mechanism capable of specifying total tensile load with precision of one percent of the specified data or better is utilized. Nowadays, digital indicators for load measurements are used which is simpler to read as compared to analog indicators. Extensometer is employed to examine the distance among two chosen points located within the gauge length of the test sample as the sample is stretched. Commercially utilized tensile testing machine is shown in the Fig. 4.2. Manual calculations are avoided and time is lessened due to newer microprocessor technology. Calculations that includes stress, modulus, elongation, statistical and energy are carried out automatically through computer.

4.3 Flexural Tests

The flexural properties determined from stress-strain characteristics are essential for the design engineer as well as plastic manufacturer. Flexural strength is used to determine the material capability towards bending forces that are subjected to perpendicular forces acting on the longitudinal direction. Combination of tensile and compressive stresses yields stresses produced through the flexural load. Flexural properties are described and computed in terms of the maximum stress and strain that occurs outside the surface of the test specimen. Ultimate flexural strength is not seen for most of the polymers and large deformation is observed under flexural load without breaking. Flexural tests offer several merits over tensile tests. If a material is acted on the geometry of beam and if service failure takes place in bending, then a flexural test is essential for preparing the design than a tensile test. One of the merits of the flexural test is that it provides the measurement of the actual deformation precisely. Two techniques (three point loading system and four point loading system) are used to conduct the flexural tests for plastics. In three point loading system (Fig. 4.3), bar of rectangular cross-section fixed on simply supported beam which is loaded using a loading nose in the centre of the bar among the supports. Loading nose provides the maximum axial stresses at the centre of the bar. In four point loading system (Fig. 4.4), even gap of the two load points from their adjacent support system is provided to conduct the tests with a distance among load points of 1/3 of the support span. In this technique, the bar specimen fixed on two supports and is subjected to load at two points with loading nose. Maximum axial stresses acts on the two loading nose for the bar specimen. Influencing factors in the flexural test results are the temperature, specimen preparation and test conditions [5].

Universal testing machine is deployed to conduct the flexural tests and also used to perform tensile and compression tests. The testing machine consists of two members; one end is fixed and other is movable member. Commonly used universal testing machine (Fig. 4.5) is provided. The machine employed for this objective should run at a constant rate of crosshead movement. Error in the measuring (load) system should not go beyond one percent of the maximum load. The support and loading nose takes the shape of cylindrical surface. The range of the nose radius must be at least one-eighth inches to eliminate unnecessary indentation or failure because of stress concentration straight under the loading nose. The deflection in the specimen is measured using strain gauge that is named as compressometer or deflectometer.

4.4 Compression Tests

Compressive properties explain the characteristic of a material while it is loaded with compression at a moderately minimum and uniform rate of loading. Even though, majority of applications of the plastics are used to subject compressive loads, still, compressive strength of plastics products has partial design value. Design considerations of plastics are performed generally using impact, fatigue and creep test results.

Fig. 4.3 Three point loading system

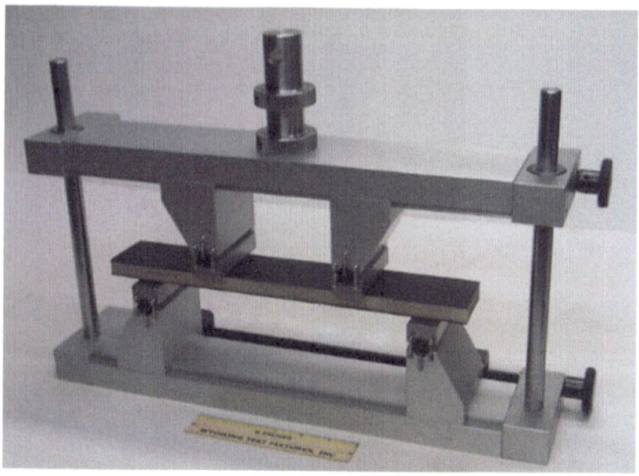

Fig. 4.4 Four point loading system

Compression tests offer a standard technique of acquiring values for quality control, research and development, rejection or acceptance under certain specifications, and special objectives. Compressive tests provide the yield stress, modulus of elasticity, plastic deformation, compressive strain, compressive strength and slenderness ratio. In most of the design guides, compressive modulus and compressive strength are the values that are of importance [6].

Compressive tests are carried out through the universal testing machine (Fig. 4.5) that may be used for conducting tensile and flexural tests. The universal testing machine requirement and operation are discussed on the tensile test section elaborately. Measurements in the change of distance are found using compressometer or deflectometer.

Fig. 4.5 Universal testing machine

4.5 Thermal Analysis Techniques

Thermal analysis (TA) comprises of analytical methods in which a property of the specimen is viewed against temperature or time while the temperature of the specimen is set. Properties contain energy take-up, dielectric constant, weight, dimension, differential temperature, evolved gases and mechanical modulus. Usage of thermal analysis is essential in the polymer industries. Thermal analyzers are adopted to identify the quality of the material for fitness of use. Thermal analysis is performed with various analyzers such as thermogravimetric analyzer, and differential calorimetric analyzer. Readers are recommended to read the books [7] for depth insight in the thermal analysis methods.

4.6 Differential Scanning Calorimetry

Differential scanning calorimetry (DSC) is adopted to identify the heat capacity and heating or cooling of the specimen. DSC (Fig. 4.6) is broadly used technique for thermal analysis. The heat flow rate to the specimen with differential power is evaluated while temperature of specimen, in a particular atmosphere, is set. Flow of heat of the specimen is examined, because all polymers have a finite heat capacity, and heating or cooling a specimen outcome. The specimen and reference sample are heated in the common chamber to find the heat flux using DSC.

The variation in the heat flow to the sample is proportional to the temperature variation that is established between specimen and reference sample of a thermocouple. Temperature around the specimen and reference sample is controlled separately by the power compensations approach. The differential energy flow is recorded using

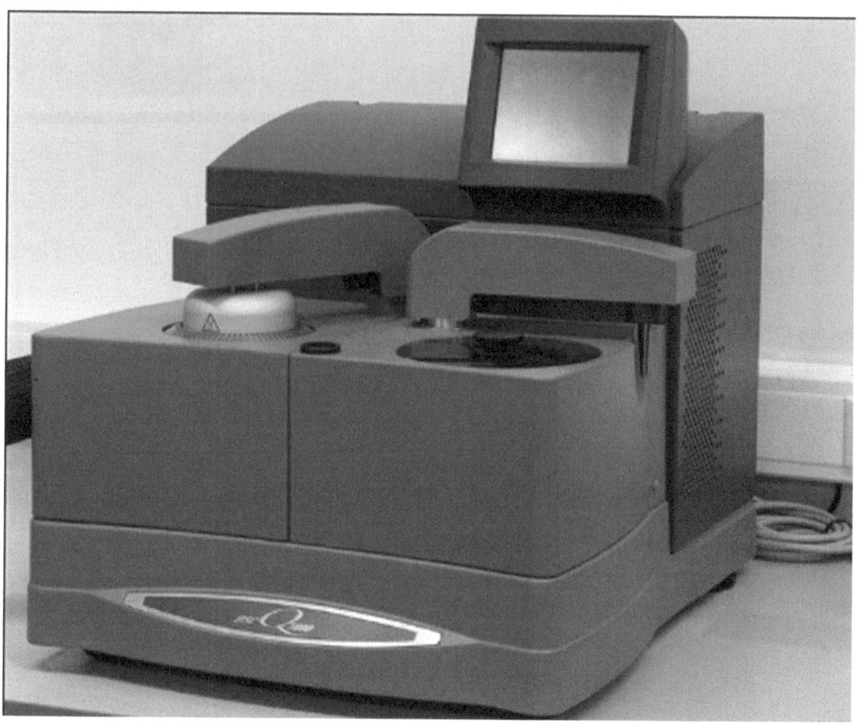

Fig. 4.6 Differential scanning calorimeter

augmented feedback from platinum resistance thermometers to keep the sample on the particular temperature. Heat flow is the output of the DSC which is denoted as watts/gram, milliwatts, or watts/gram-degree.

4.7 Thermogravimetric Analysis (TGA)

Thermogravimetric analysis (TGA) is performed to examine the mass loss of the specimen during progressive heating. Temperature is increased either at a constant rate or by a series of steps to identify the mass loss of the specimen. Decomposition of the polymer or elastomer materials is observed through thermogravimetric analysis. Quantitative measurements are carried out during series of weight-loss steps. Distinctive high-performance equipment comprises of an analytical stability supporting a platinum crucible for the work piece, the former placed in a furnace. Differences in instrumentation contain horizontally placed furnaces and balances. Thermogravimetric analyzer is shown in the Fig. 4.7.

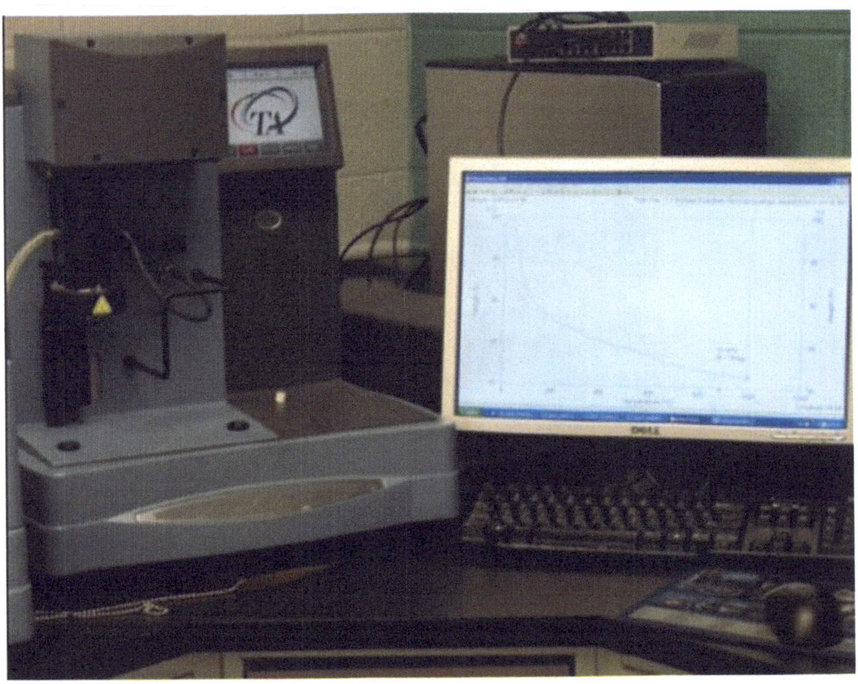

Fig. 4.7 Thermogravimetric analyzer

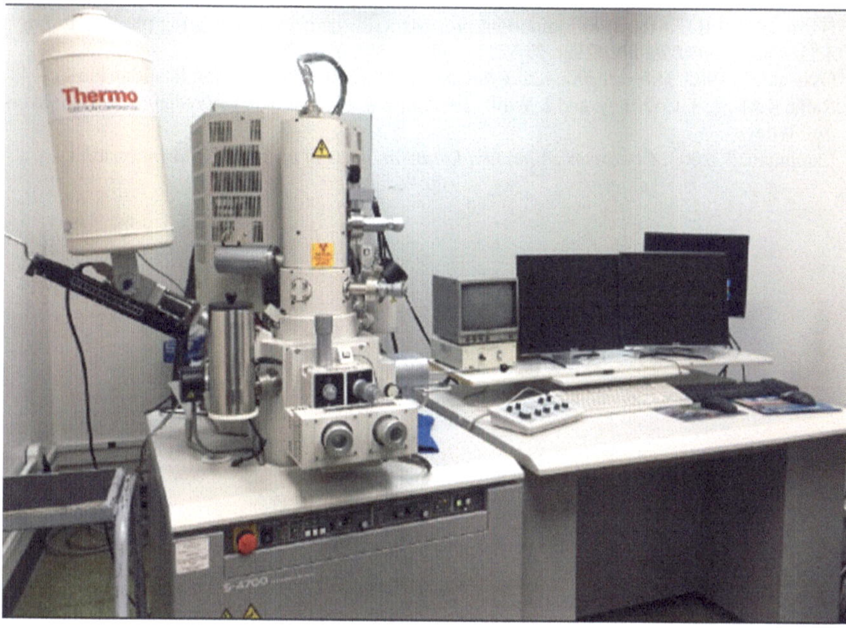

Fig. 4.8 Scanning electron microscopy for polymer welding

4.8 Scanning Electron Microscope

Plastics industry relies on electron microscope to investigate the size of the particles, the thickness of the foils and the diameters of fibers. These results assist in enhancing the polymer composition and manufacturing process, thereby improving the performance of any manufactured goods. Electron beam is directed on the specimen with higher magnification in the scanning electron microscope (Fig. 4.8). Electron beam is focused on the specimen at the vacuum conditions to avoid the interaction with unwanted contaminants in the air. Secondary electrons are discharged from the specimen during hitting of electron beam on the specimen which provides the topolographical image of the specimen surface. Secondary electron detector (SED) and backscattered electron detector (BSE) are generally used as detectors in the scanning electron microscope. Image is created during electrons interact. Special preparations are essential to accomplish the microstructures of the polymer materials.

References

1. Harper, C.A. 2000. *Modern plastics encyclopedia*. McGraw Hill.
2. Shah, V. 1998. *Handbook of plastics testing technology*, 2nd ed. Wiley.
3. ISO/IEC. 1999. *Selected standards for testing plastics*, 2nd ed. ASTM.

4. Nairn, J., and R. Farris. 1988. Important properties divergences, engineering plastics. In *Engineered materials handbook,* vol. 2. ASM International, 655–658.
5. Osswald, T. 1998. *Polymer processing fundamentals.* Hanser/Gardner Publications Inc., 19–43.
6. Ralls, K.M., T.H. Courtney, and J. Wulff. 1976. *Introduction to materials science and engineering*. Wiley.
7. Lampman, S. 2003. *Characterization and failure analysis of plastics*. ASM International.

Chapter 5
Data Acquisition and Optimization Techniques for USW

5.1 Data Acquisition Systems

Data acquisition (DAQ) is the key sequence in measurement of physical or an electrical phenomenon such as temperature, pressure, sound, voltage, or current using a computer. DAQ system (Fig. 5.1) comprises various sensors for appropriate applications, and the DAQ hardware is integrated with software to be viewed on computer. Compared to conventional measuring devices, computer-based DAQ devices employ the productivity, processing power, display, and connectivity abilities of industry standard computers offering more flexible, powerful, and inexpensive measurement solution.

5.1.1 Physical Phenomenon of DAQ

5.1.1.1 Sensors

Physical parameter measurements such as the intensity of a light source, temperature of a room or the force applied to an object start with a sensor. Physical parameter is recorded into a measurable electrical signal by a transducer. Electrical output can be in terms of change in resistance, current, voltage or any other electrical signal that differs over time based on the nature of sensor. Few sensors may need further components and circuitry to generate a signal that can precisely and safely be acquired by a DAQ device.

S. A. Vendan et al., *Confluence of Multidisciplinary Sciences for Polymer Joining*,
https://doi.org/10.1007/978-981-13-0626-6_5

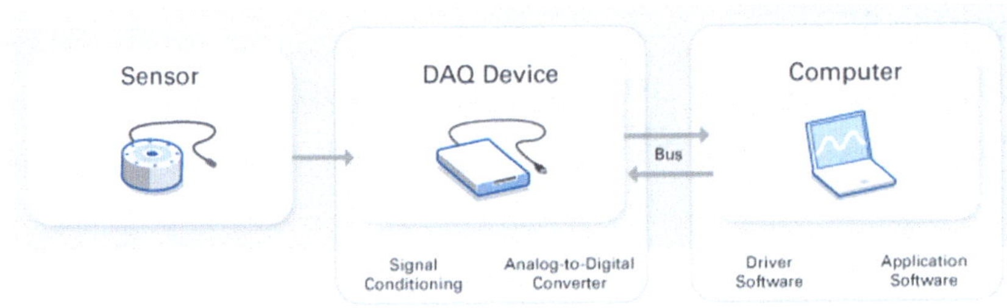

Fig. 5.1 Components of the DAQ system [*Courtesy* National Instruments]

5.1.1.2 DAQ Device

Signals from the other machines are acquired by the DAQ hardware, and it is connected to computer to interpret results. DAQ hardware plays the role of interface among the computer and the machines to acquire results. Analog signals are converted into digital output using DAQ device. Signal measurement device, that is DAQ, comprises of three major components such as analog-to-digital converter (ADC), signal conditioning circuitry, and computer bus. Some other DAQ devices incorporate other purposes such as digital I/O lines input and output digital signals, digital-to-analog converters (DACs), output analog signals, and counter/times count and produce digital pulses.

5.1.1.3 DAQ Device Components

Signal Conditioning

Signal while measuring using sensors is converted into a form that is appropriate for input into an ADC influenced by signal conditioning circuitry. This circuitry can perform the attenuation, amplification, isolation, and filtering. Specific types of sensors measure through some DAQ devices designed with built-in signal conditioning.

Analog-to-Digital Converter (ADC)

Analog signals measured are converted into digital signals for better signal quality and back to analog post-transmitting by ADC and DAQ. Digital representation of an analog signal is performed through the ADC chip. Continuous variation in the analog signals with time is observed, and an ADC obtains intermittent samples of the signal at a predefined rate. These samples are relocated to a computer over a computer bus where the unique signal is recreated from the samples in software.

Computer Bus

Computer using a port or slot is linked with DAQ devices. The computer bus delivers the data through the communication device between the computer and DAQ device for measuring data and passing instructions. DAQ devices are provided on the most general computer buses containing PCI Express, USB, PCI, and Ethernet. Nowadays, wireless DAQ devices are available. There are numerous kinds of buses and each provides various merits for various categories of usages.

Computer's Role in DAQ system

Operation of the DAQ device is controlled through computer with programmable software, and computer is utilized for visualizing, processing, and storing measurement data. Various categories of computers are used based on the specific applications. Industrial computers employed in production unit are attributed to its roughness; laptop may be utilized for its mobility; or a desktop may be used in a laboratory for its processing power.

Application Software

Interaction among the computer and observer is facilitated through the application software that is used for analyzing, acquiring, and providing measurement data. Application is built with predefined functionality or custom functionality based on the requirement. Automation of the various purposes of a DAQ device is carried out for specific applications, and these applications conduct signal processing through well-defined algorithms and display the desired results with the help of user interfaces [1].

5.1.1.4 Acquisition of Temperature Using USW

One of the significant parameters in the ultrasonic welding is temperature at the interface of the specimens. Hence, it is measured at various points through DAQ system. Temperatures at the interface of the specimens are collected using DAQ system with K-type (−270 to 1260 °C) thermocouple sensor, DAQ card, and LabVIEW software. The block diagram of data acquisition system of the ultrasonic welding setup is provided in Fig. 5.2. DAQ card fetches the signal from the K-type thermocouple. Figure 5.3 is provided in the experimental setup of the ultrasonic welding with DAQ system. Programming of the data fetching for a certain duration is carried out through LabVIEW software.

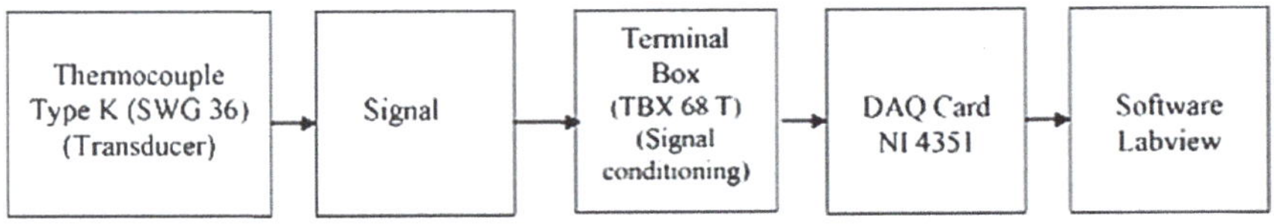

Fig. 5.2 Process of data acquisition system

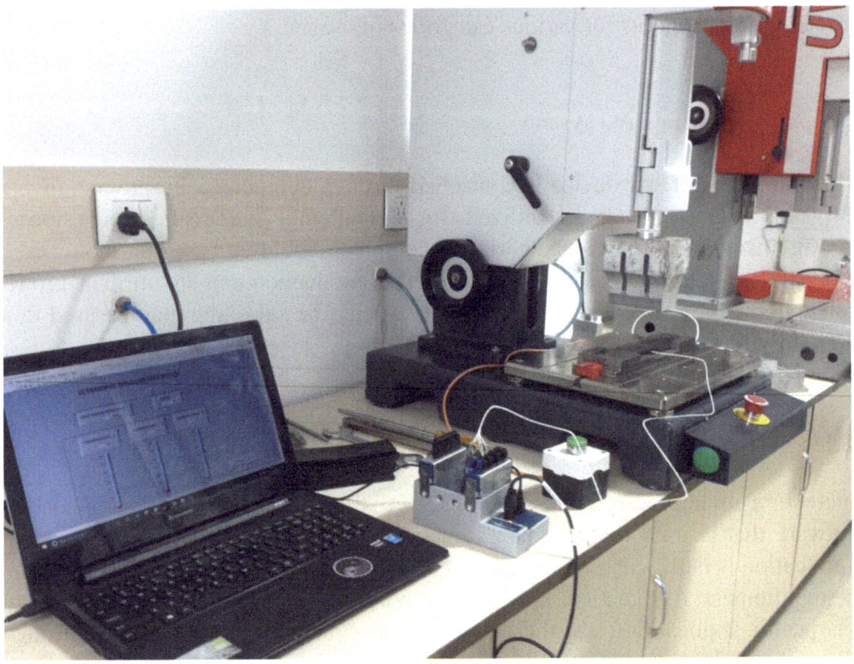

Fig. 5.3 Experimental setup for temperature measurement

5.2 Control System for USW

Desired response is achieved through the control system that links different system components. The control theory indicates input–output (cause–effect) relationship for various parameters involved in the USW system. Three main tasks of control system are

- Measurement of output parameter;
- Comparison of measured output to a reference value;
- Self-adjustment mechanism to minimize the error which is the difference between measured output and a reference value.

When some elements are connected in a series to carry out a specific function, it is called as system. For example, an USW setup (Fig. 5.4) is a system comprising

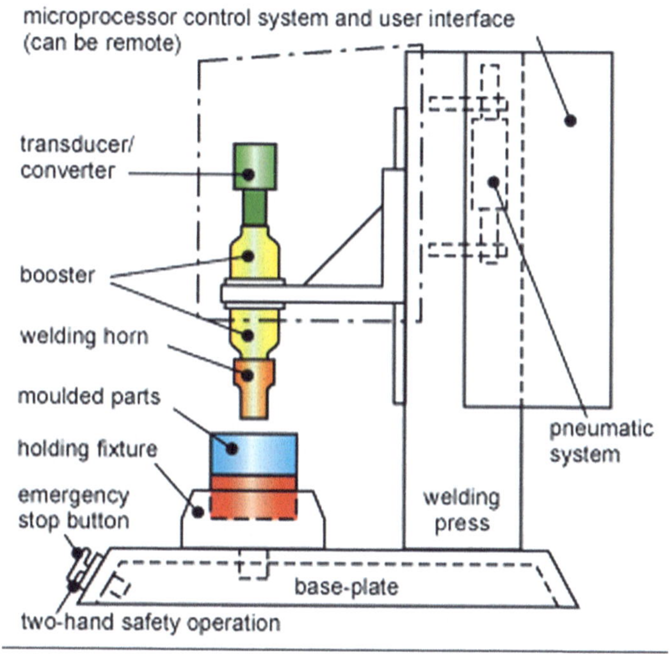

Fig. 5.4 Schematic diagram of ultrasonic welding machine

various elements like power source, piezoelectric transducer, sonotrode, workpieces, clamps, holders, and sensors connector in series.

Order of the system refers to the uppermost derivative of the controlled quantity in a mathematical equation explaining the dynamics of the system, i.e., the highest power of 's' in the denominator polynomial of the transfer function without terminating the common terms in the numerator and denominator polynomial.

- A zero-order system is one in which output changes instantaneously as the input changes. It is a memoryless system.
- If the input–output equation of a system is a first-order differential equation, it is called as first-order system. For example, thermocouple is an example of first-order system. Measurement of temperature distribution in the heat-affected zones (HAZ) of the polymer sheet joined by USW process can be performed using a thermocouple [2].
- The second-order system is represented by the differential equation as shown in the following equation. There are several factors that make second-order systems important [3].

$$\frac{C(s)}{R(s)} = \frac{\omega_n^2}{s^2 + 2\xi\omega_n s + \omega_n^2}$$

Table 5.1 Basic control actions

Control action	Mathematical notation	Effects
Proportional control	$u(t) = k_p e(t)$	Adjustable gain (amplifier)
Integral control	$u(t) = k_i \int_0^t e(t)dt$	Eliminates offset (deviation in the response of the system from the desired settling value) error. Also causes oscillations
Derivative control	$u(t) = k_d \frac{de}{dt}$	Provides faster response. Never used alone

where
$u(t)$ control signal
$e(t)$ control error ($e =$ feedback signal − setpoint value)
k_p, k_i, k_d are the proportional gain, integral gain, and derivative gain, respectively

where

ω_n natural frequency (rad/s)
ξ damping ratio.

Two categories of analysis (time domain analysis and frequency domain analysis) may be used to express the results of a control system. Function of time is considered from the system response in time domain analysis. In addition, major components of time domain analysis are transient response and steady-state response. The time domain analysis can be employed only after modeling the system mathematically and knowing the input of the system. Welding applications typically use time domain analysis. Function of frequency is taken from the system response in frequency domain analysis. Phase response and magnitude response are the components of frequency domain. Physical systems without finding its mathematical model are used to measure the values of significant feature of frequency response.

Fixed structure controller family or proportional, integral, and derivative (PID) controller family is the fundamental controllers. These controllers are reliable and can be used in the control of key process parameters to achieve good weld. The basic control actions are shown in Table 5.1 [3].

In common, PID controller is adopted in most of the closed-loop industrial processes in spite of the large quantity of sophisticated controllers. The general equation of a PID control is represented by Moré [3].

$$u(t) = k_p e(t) + k_i \int_0^t e(t)dt + k_d \frac{de}{dt}$$

The important function of a feedback control system is to make sure that the closed-loop system has a stable dynamic and steady-state response behavior [4]. Preferably, the closed-loop system should satisfy the following performance criteria:

(i) Stability and robustness;

(ii) Minimal disturbance providing good disturbance rejection;
(iii) Good set point tracking (i.e., rapid smooth responses to set point changes);
(iv) Steady-state error (offset) is eliminated;
(v) Avoiding excessive control action.

All the ultrasonic welding systems come with their inbuilt controllers within the machine.

5.3 Optimization Using Artificial Neural Networks for USW

Artificial neural networks (Fig. 5.5) are simplified model of biological network structure. Artificial neuron is the brain (building block) of the ANN. Receiving and transmitting data are carried out through the interactions of neurons with the real world. Hidden layers comprise several hidden neurons. Neurons are linked to one another through synapse, and each synapse is connected with a weight factor. Artificial neural network (ANN) has their obscure role in real time, interactive, and complex usages such as pattern recognition, speech recognition, finance, medicine, weather forecasting, and control of production methods. The favorable characteristics of ANN are that they can be modeled through the experimental values. Various models are employed in the ANN algorithms such as multilayer perceptron (MLP), back-propagation (BPN), radial basis function (RBF), self-organizing map (SOP). [5, 6].

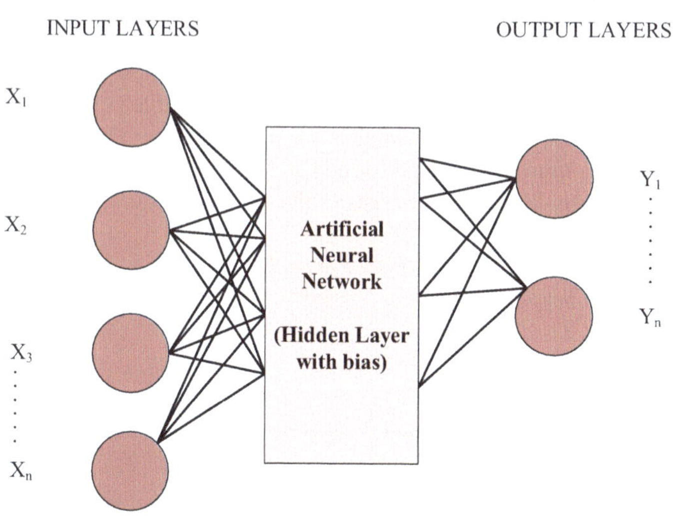

Fig. 5.5 Artificial neural network model

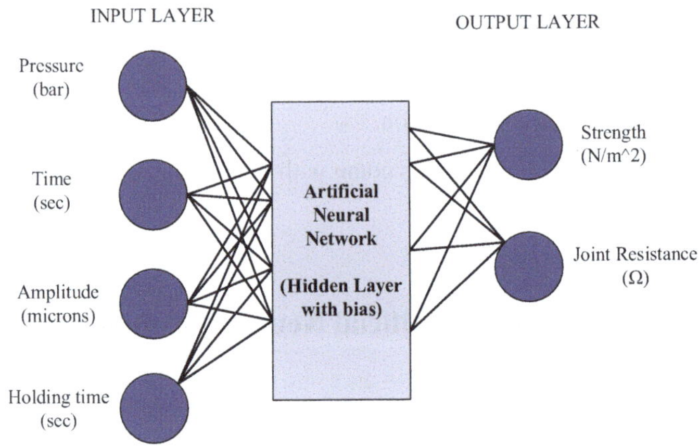

Fig. 5.6 ANN model of USW

For USW process, the ANN model can be developed with four set of input parameters: pressure (bar), weld time (sec), amplitude (microns), and holding time (sec). The hidden layers are chosen in the implemented algorithm. The output layer can be of any desired parameter (as shown in Fig. 5.6). Several data samples are used for training the ANN model, and few of the samples may be used for prediction and testing.

5.3.1 Introduction to LM Algorithm

In this investigation, Levenberg–Marquardt (LM) optimization is used to train the ANN. LM is a virtual standard in nonlinear optimization which carries out gradient descent and conjugate gradient techniques for the problems. The function evaluations and gradient information are performed and estimated through Hessian matrix. Basically, Levenberg–Marquardt consists of the following equation for solving,

$$\left(J^T J + \lambda I\right)\delta = J^T E$$

where

- J Jacobian matrix for the system
- λ Levenberg's damping factor
- δ weight update vector
- E error vector.

The LM algorithm is designed to lessen the sum of squared error function of the form [3],

$$E = \frac{1}{2}\sum k(e_k^2) = \frac{1}{2}\|e\|^2$$

where

e_k is the error.
e is a vector with element e_k.

Based on δ, the network weights are changed to achieve appropriate solution. The $J^T J$ matrix is known as the approximated Hessian. The damping factor is adjusted in each iteration.

5.3.2 Computing the Jacobian Matrix

The Jacobian matrix consists of all first-order partial derivatives of a vector function. In the neural network, it is a N × W matrix, where **N** is the number of elements in training set and W is the total number of parameters (weights + bias). It is created by taking the partial derivatives of each output with respect to each weight denoted by the form [7],

$$J = \begin{bmatrix} \frac{\partial F(x_1,w)}{\partial w_1} & \cdots & \frac{\partial F(x_1,w)}{\partial w_W} \\ \vdots & \ddots & \vdots \\ \frac{\partial F(x_N,w)}{\partial w_1} & \cdots & \frac{\partial F(x_N,w)}{\partial w_W} \end{bmatrix}$$

where $F(x_1, w)$ is the network function evaluated at ith input vector of training set using the weight vector w and w_W is the wth element of the weight vector w.

5.3.3 Steps in Levenberg–Marquardt Algorithm

1. The LM involves solving the equation of $J^T E$ with different values of λ values until the sum of squared error decreases. Each iteration involves following basic steps:
2. Computation of the Jacobian matrix.
3. Computation of error gradient $g = J^T E$.
4. Approximation of Hessian using the cross-product Jacobian $H = J^T J$.

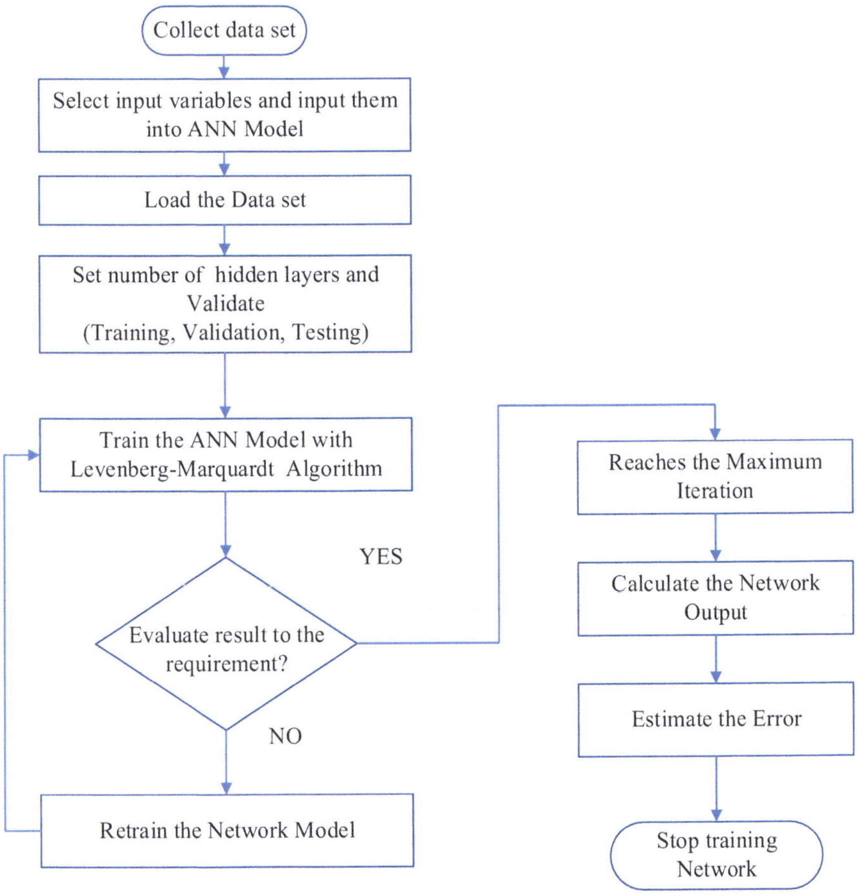

Fig. 5.7 Flowchart of the LM-based ANN model

5. Solving the equation: $(H + \lambda I)\delta = g$ find δ.
6. Updating of the network weights w using δ.
7. Recalculation of the sum of squared errors using updated weights.
8. If the sum of squared errors not decreased, discard the new weights, increase λ using some adjustment factor and go to step 4.
9. Else decrease λ and stop.

The flowchart of the implemented ANN model is depicted in Fig. 5.7.

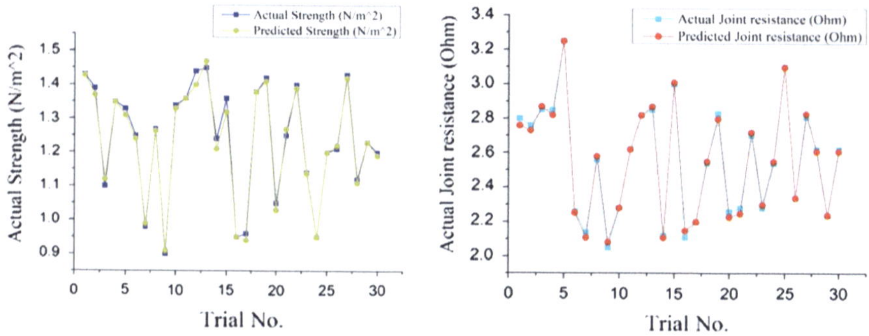

Fig. 5.8 Prediction of strength (left) and joint resistance (right)

Table 5.2 Testing and validation of evaluated parameters

Test no	Actual strength (N/m^2)	Predicted strength (N/m^2)	Error (%) for strength	Actual joint resistance (Ω)	Predicted joint resistance (Ω)	Error for joint resistance (%)
1	0.8	0.81	−1.10	2.15	2.073	−1.5
2	1.3	1.33	0.72	2.18	2.271	−0.03
3	1.32	1.36	0	2.42	2.61	0

5.3.4 *Prediction of Strength and Joint Resistance Using LM Algorithm-Based ANN*

Based on the experimental data, the ANN is trained using LM algorithm in the MATLAB tool platform to predict the strength and joint resistance of ultrasonically joined polymer sheets (polycarbonate). Figure 5.8 shows that the strength and joint resistance predicted by the developed LM-based ANN model which shows good coherence with the experimentally measured values.

5.3.5 *Testing and Validation*

Three test samples are taken for validating experimentally recorded results. The test results reveal that the experimental and predicted values are in close coherence with each other as shown sin Table 5.2.

The percentage difference of error while measuring the spread of prediction by the ANN model (in predicting the strength and joint resistance) is estimated. It is observed that the developed LM-ANN predicts within the acceptable error limits. The error plots are shown in Fig. 5.9.

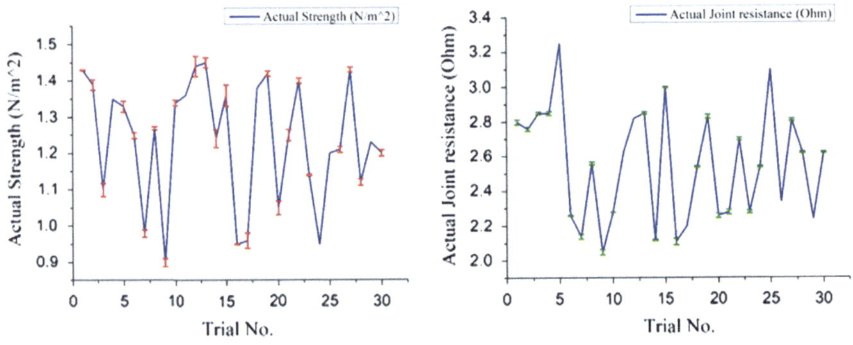

Fig. 5.9 Error plot in prediction of strength (left) and joint resistance (right)

References

1. Manual, L.U. 1998. *National instruments*. Austin, TX.
2. Suresh, K.S., M.R. Rani, K. Prakasan, and R. Rudramoorthy. 2007. Modeling of temperature distribution in ultrasonic welding of thermoplastics for various joint designs. *Journal of Materials Processing Technology* 186 (1–3): 138–146.
3. Moré, J.J. 1978. The Levenberg-Marquardt algorithm: Implementation and theory. In *Numerical analysis*, 105–116. Berlin, Heidelberg: Springer.
4. Peters, S.R., and Fulmer, B.E. 2015. *Non-linear adaptive control system and method for welding*. Lincoln Global Inc., U.S. Patent 8,963,045.
5. Mandal, S., P.V. Sivaprasad, S. Venugopal, and K.P.N. Murthy. 2009. Artificial neural network modeling to evaluate and predict the deformation behavior of stainless steel type AISI 304L during hot torsion. *Applied Soft Computing Journal* 9 (1): 237–244.
6. Vasudevan, M. 2009. Soft computing techniques in stainless steel welding. *Materials and Manufacturing Processess* 24 (2): 209–218.
7. Gale, D., and H. Nikaido. 1965. The Jacobian matrix and global univalence of mappings. *Mathematische Annalen* 159 (2): 81–93.

Chapter 6
Case Studies on Ultrasonically Welded Polymer Joints

6.1 Joining of PC (65%) + ABS (35%) Blend to PC (65%) + ABS (35%) Blend Using Ultrasonic Welding Method

6.1.1 Research Task

Amorphous (PC + ABS) polymers are welded using ultrasonic welding process which is the aim of this investigation. Polymer granules are molded into rectangular plates with energy director embedded onto it before welding of polymer specimens. Taguchi method is employed to design the welding parameters in this investigation. Differential scanning calorimetric (DSC) and thermogravimetric analysis (TGA) are performed to examine the thermal behavior of the welded specimens. Weld strength of the polymers is determined using the mechanical testing. Finite element and microscopical analysis are also reported in this study. In the end, RSM and ANOVA techniques are applied and presented.

6.1.2 Problem Formulation

Ultrasonic welding of polymers involves the following steps such as synergistic friction movement, heat propagation, smoothening of the material and diffusion under the elevated temperature, flow of the material and solidification. Thus, depth insight is crucial for the ultrasonic thermomechanical welding methods, and a well-defined methodology is required to deal with polymer joining due to its complexity. Yet, experimental results reveal the contradictory and conflicting inferences from the previous literature on ultrasonic welding.

Weldability and unidirectional coupled result of ultrasonic weld formation or parameter optimization are the behavior discussed mechanically for the ultrasonic

S. A. Vendan et al., *Confluence of Multidisciplinary Sciences for Polymer Joining*,
https://doi.org/10.1007/978-981-13-0626-6_6

welding models. These models are yet unproductive to realize the physical and thermomechanical joining phenomena. Mechanical and thermal analysis is carried out without any direct association. Correspondingly, modeling work validation is discussed in the literature owing to the complexities of mechanical, thermal, and displacement measurement.

This research reveals the thermal and mechanical characteristics of the polymer welding using ultrasonic welding process. This provides an insight thermomechanical welding at the interface of the polymer materials. The energy director is embedded, and the impact of mechanical and thermal deformation is assessed. The proposal of weld strength determination and weld strength improvement is suggested. The contributions and functions of mechanical and thermal process parameters to the ultrasonic welding are determined and used in experimental trials. This research also records the microscopic analysis and finite element analysis for depth insight of weld phenomenon.

6.1.3 Introduction to Materials and Methods

This section reports the intricate information on the materials used for this research that encloses their mechanical, chemical, and structural properties. The manufacturing method entailed in mold preparation and fabrication is illustrated elaborately. The process parameter selection is vital to accomplish proper welding, and the parameters governing the process are amplitude of vibration, pressure, hold and weld time, energy director (ED), and materials. In the previous section, sufficient data are available for ultrasonic welding related to polymers reported which lead the way for this study. It is essential reason necessary to realize the material performance related to their property modifications that depend on process parameters.

6.1.3.1 Materials

The materials selected for this investigation are polycarbonate (PC) +acrylonitrile butadiene styrene (ABS) blend. Granules of polycarbonate and acrylonitrile butadiene styrene are shown in Figs. 6.1 and 6.3. Chemical structure of the materials is shown in Figs. 6.2, 6.4, and 6.5. Properties of the materials are listed in Tables 6.1, 6.2, and 6.3.

6.1.3.2 Manufacturing Process

The injection molding methods to shape the polymer materials using mold surface or cavity and ultrasonic welding process of PC + ABS is discussed in this section.

Fig. 6.1 Polycarbonate granules

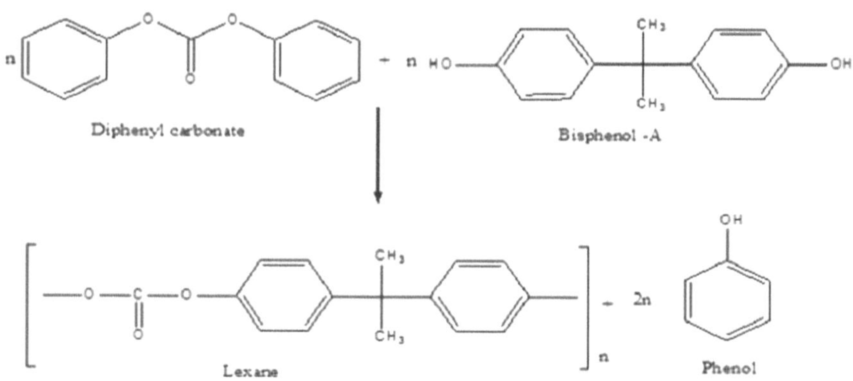

Fig. 6.2 Chemical structure of polycarbonate

6.1.3.3 Injection Molding

Preparation of energy directors: The die is designed and fabricated with EDs in the injection molding method. The triangular shape of the energy director is shown in Fig. 6.6. The part geometry, shape of sonotrode, and part material are the parameters used to design the energy director. The maximum energy is absorbed by the energy director during ultrasonic welding of thermoplastics.

Injection molding: It is used to pack the molten material into the mold cavity using the pressure. The solidification of the molten material occurs at the lower melting point and produced the final product. Various stages are involved in the

Table 6.1 Properties of polycarbonate

Properties	Parameters	Values
Physical	Density (g/cm^3)	1.2
Mechanical	Tensile strength (psi)	9500
	Flexural strength (psi)	15,000
	Compressive strength (psi)	12,000
Thermal	Heat deflection temperature (°C)	132
	Glass transition temperature (°C)	145
	Max. operating temperature (°C)	151
	Thermal conductivity (°C)	6.9

Table 6.2 Properties of ABS

Properties	Parameters	Values
Physical	Density (g/cm^3)	1.05
Mechanical	Tensile strength (psi)	7750
	Flexural strength (psi)	11,800
	Compressive strength (psi)	–
Thermal	Heat deflection temperature (°C)	94
	Glass transition temperature (°C)	105
	Max. operating temperature (°C)	105
	Thermal conductivity (°C)	−16

Table 6.3 Properties of PC + ABS

Properties	Parameters	Values
Physical	Density (g/cm^3)	1.098
Mechanical	Tensile strength (psi)	5900
	Flexural strength (psi)	9800
Thermal	Heat deflection temperature (°C)	110
	Glass transition temperature (°C)	125

Fig. 6.3 ABS granules

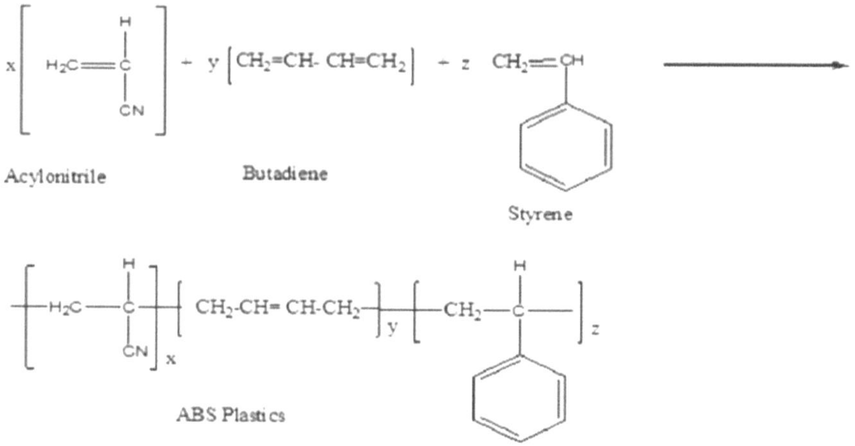

Fig. 6.4 Chemical structure of ABS

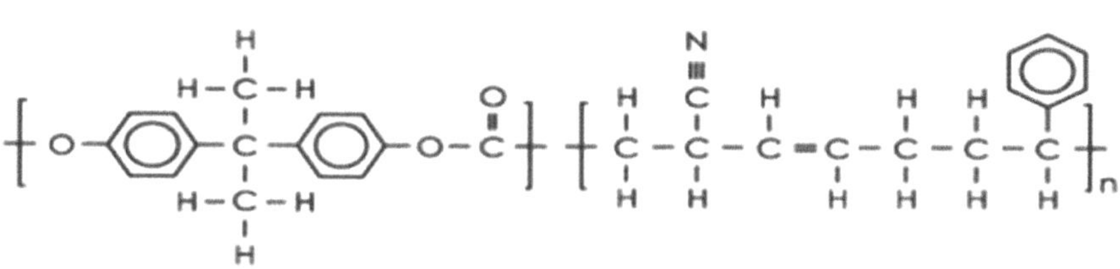

Fig. 6.5 PC + ABS composites structure

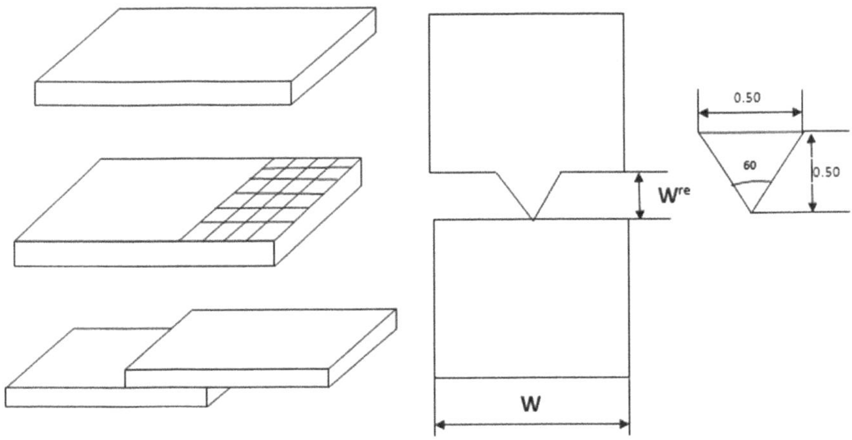

Fig. 6.6 Schematic of triangular energy director

Fig. 6.7 Mold setup

injection molding method such as plasticity phase, melting, homogenizing phase, and injection phase. ABS 35% and PC 65% are used in this study.

The illustration of the die setup is depicted in Fig. 6.7, and the tabulation of the injection molding process parameter settings is recorded in Table 6.4. The high holding pressure is utilized to balance the thermal variation at the phase of cooling. After the solidification, the molded specimen is taken and is subjected to ultrasonic welding.

6.1.3.4 Ultrasonic Polymer Welding Process

Ultrasonic welding machine comprising of 20 kHz power levels from 2000 to 2400 W is used for this investigation for joining the PC + ABS blend components as shown in Fig. 6.8. The higher characteristics such as patented trigger by power apart from

Table 6.4 Injection molding parameters

S.No.	Parameters	Values
1	Injection time (s)	5
2	Holding time (s)	5
3	Injection pressure (kPa)	230
4	Holding pressure (kPa)	175
5	Cooling time (s)	35
6	Injection temperature	
	• Nozzle zone (°C)	175
	• Front zone (°C)	180
	• Center zone (°C)	185
	• Rear zone (°C)	190

Table 6.5 Parameter levels od PC + ABS blend welding trials

Factors	Units	Levels		
		Level 1	Level 2	Level 3
Pressure (P)	Bar	3	3.5	4
Weld time (T)	S	0.5	1.0	1.5
Amplitude (A)	μm	20	30	40

time, process limits, distance controllable provisions, and energy are provided by ultrasonic welding machine. The ultrasonic vibrations to produce the weld use higher pressure, constant frequency (20 kHz), and low amplitude (20–40 μm). The joining of polymers occurs due to the heat produced due to the stresses, intermolecular friction and continuous forced contact. In this ultrasonic joining of polymers, heat generation is the governing parameter. The energy director design (EDs) process parameter selection is the cause for melting of polymer.

The pressure, weld time, and amplitude are major parameters influencing the weld strength and quality. Further, these parameters are designed with design of experiments (DOE) using three different levels. The material characteristics are the key parameter for the welding of amorphous and semicrystalline materials which have diverse melting temperature and properties. Weld area, energy director (Fig. 6.9) positions, and schematic of lap joint are shown in figure. Parameter levels used in PC + ABS blend are listed in Table 6.5. Consequently, the Taguchi method (L27) is adopted to conduct experimental trials. Figure 6.10 shows the welded specimens of PC + ABS blend.

Fig. 6.8 Ultrasonic welding equipment

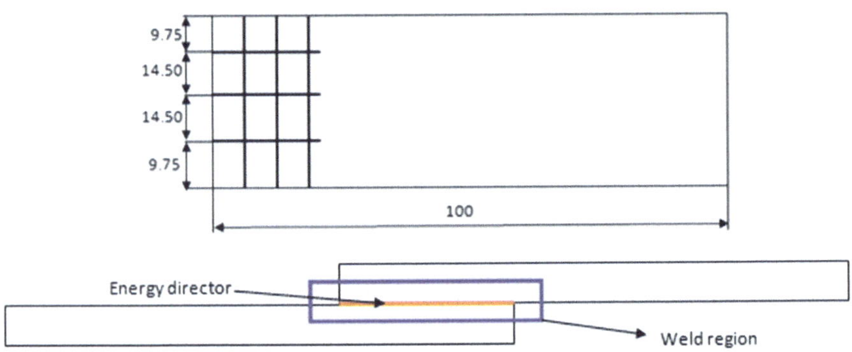

Fig. 6.9 Lap joint design incorporated with energy director

6.1.4 Results and Discussion

The welded specimens are analyzed with various methods for the mechanical, thermal, and integrity analysis.

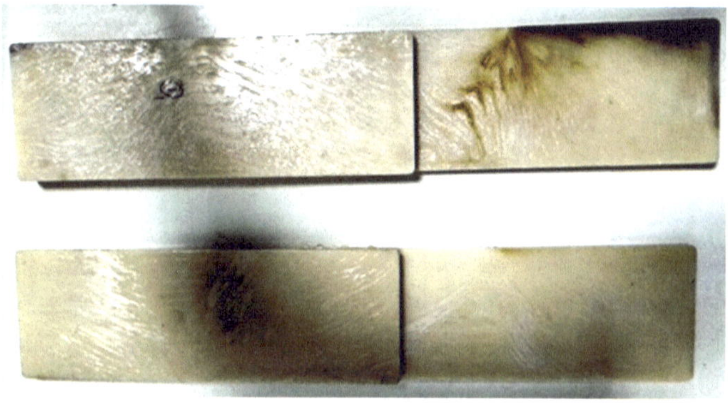

Fig. 6.10 Ultrasonically welded PC + ABS blend

6.1.4.1 Mechanical Analysis

The weld strength is determined by the mechanical strength of the welded specimens. The various categories of evaluation techniques are available to characterize the weld strength. The different techniques adopted to test the welded specimens are tensile strength, bending test, ultrasonic test, peel test, and scratch test. The joint is subjected to the forces in perpendicular direction to the weld interface till the failure occurs during tensile testing.

Tensile Test

The various specimens of PC + ABS blend are examined by their true stress–strain curves shown in Fig. 6.11. Strain hardening effect is not found for the PC + ABS alloy at the weld interface. The strain rate of the base material is high compared with the weld specimen. The various levels of the welded specimen are determined with stress–strain curve. The curve depicts insignificant alterations with varying levels. It is proof that the welded components have considerable alteration in material properties wherein still retains majority of the original properties from the results. The observations show that the welding of the PC + ABS blend has good thermal properties and performance.

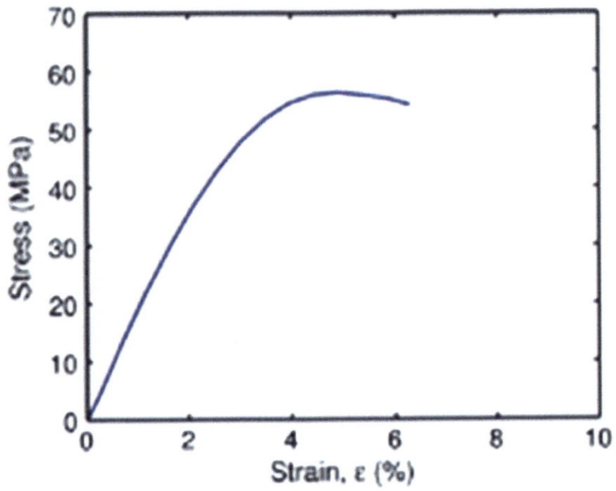

Fig. 6.11 Stress and strain curves for PC/ABS blend

6.1.4.2 Characterization Techniques

Differential Scanning Calorimetric Analysis

PC + ABS Welded Material Analysis

Differential scanning calorimetric analysis is carried out to identify the phase transition of the ultrasonically welded PC + ABS blend. The welding of the PC + ABS blend is carried out above the glass transition temperature and lesser than the decomposition temperature. DSC analysis is performed for the welded PC + ABS blend (Fig. 6.12). Six different specimens (trial and error method) are taken to observe the variation in the results. The average onset value (75.2 °C), mid value (117 °C), and end value (153.7 °C) are obtained for six different welded specimens. Variation in the Tg curve of DSC results is observed for the welded specimens due to secondary heating during welding which is lower than the molding temperature. Thermal characteristics of polymer blend are mainly affected by the weld strength. Brittle nature is seen during cooling due to the material hardness. Aforesaid may be due to the noticeable diminution in the size of the grains with respect to the vibrational heat during ultrasonic welding process.

Thermogravimetric Analysis

PC + ABS Analysis

The polymer blend with the ratio of 35% of ABS and 65% of PC specimens is subjected to analyze the thermal degradation of the welded blend specimens (Fig. 6.13). The degradation starts with a mass loss of −12.06% and initiates at 364 °C during the formation of butadiene monomer occurs. Styrene degradation starts at 460 °C,

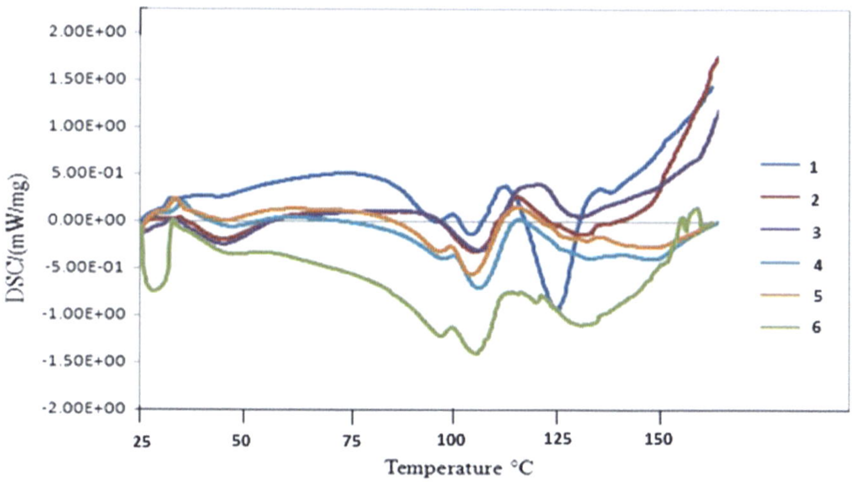

Fig. 6.12 Welded PC + ABS material DSC (5 K/min) analysis

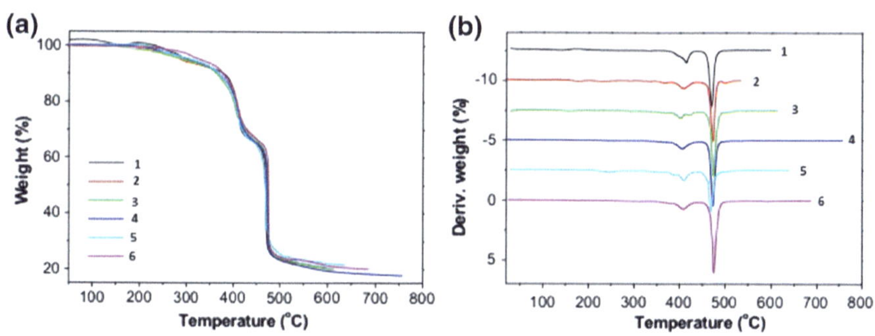

Fig. 6.13 TG and DGA results for welded PC + ABS material

and mass change is observed with 12.72% during increase in the temperature. During the onset temperature (427–490 °C), oxygen reacts with ultrasonically welded PC + ABS blend and initiates degradation by involving in the depolymerization that leads to the formation of the hydroperoxides which define a stable characteristic at a specific temperature. The degradation temperature of the PC + ABS blend is in between the degradation temperature of PC and ABS, and initially, the ABS degrades due to its lower thermal stability. Thermal degradation is lesser in the ultrasonic welding process while compared to other welding processes such as hot tool, hot gas, and resistance welding where full melting of the part occurs.

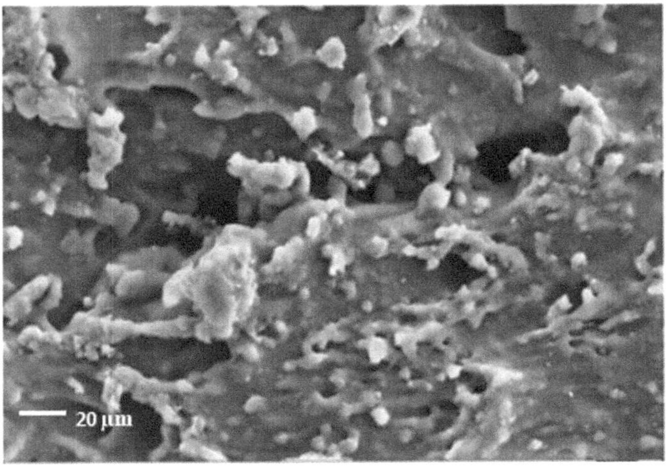

Fig. 6.14 SEM images for PC + ABS material (level 1)

6.1.4.3 Integrity Analysis

SEM Analysis for PC + ABS Weldments

Scanning electron microscopic images of the PC + ABS blends are shown in Figs. 6.14, 6.15, and 6.16. The joint interface integrity and surfaces of the welded specimens are examined through scanning electron microscopy (SEM). Several voids are observed for higher vibrational amplitude (40 μm) as compared to the 20–30 μm amplitude ranges. Nevertheless, the existence of voids has lesser influence in the lap joint interface strength, and for amplitude range of 40 μm vibrations, higher bond integrity and weld strength are recorded.

6.1.4.4 Conclusions

This investigation focuses on examining the feasibility of ultrasonic lap joining for a polymer blend including PC (65%) + ABS (35%) thermoplastics. Energy director imprinted on the weld specimens of polypropylene is made with injection molding process. The ultrasonic energy adopting ultrasonic welding process is employed to give the pressure and vibrations to the specimens for joining the specimens (PC (65%) + ABS (35%) blend). Most of the heat is transmitted along with even distribution on the weld samples.

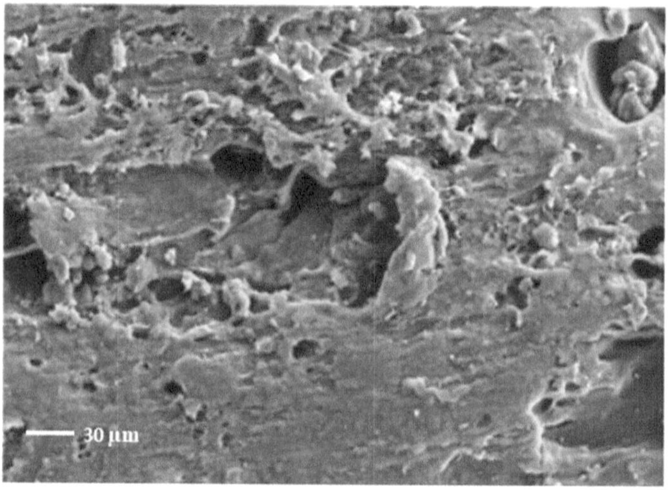

Fig. 6.15 SEM images for PC + ABS material (level 2)

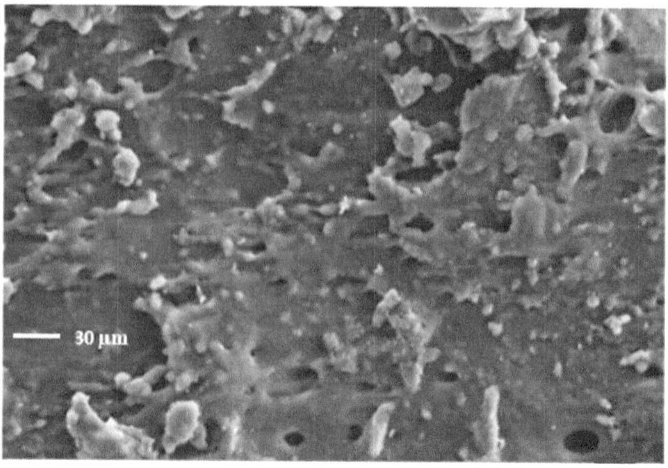

Fig. 6.16 SEM images for PC + ABS material (level 3)

6.2 Ultrasonically Joined Polypropylene

6.2.1 Research Task

Semicrystalline (polypropylene) polymers are welded using ultrasonic welding process which is the primary intention of this study. Polymer granules molded rectangular samples are embedded energy director on to the polymer specimens prior to welding. Taguchi method is employed to design the welding parameters in this inves-

Fig. 6.17 Polypropylene granules

tigation. Differential scanning calorimetric (DSC) and thermogravimetric analysis (TGA) are performed to analyze the thermal behavior of the welded specimens. Weld strength of the polymers is determined using the mechanical testing. Finite element and microscopical analysis are also reported in this study.

6.2.2 Materials and Methods

This section reports the intricate information on the materials used for this research that comprises of their mechanical, chemical, and structural properties. The manufacturing method entailed in mold preparation, and fabrication is illustrated elaborately. The process parameter selection is vital to accomplish proper welding, and the parameters governing the process are amplitude of vibration, pressure, hold and weld time, energy director (ED), and materials. In the previous section, sufficient data are presented for ultrasonic welding related to polymers paving way for this study. It is hence essential to realize the material performances related to their property modifications that depend on process parameters.

6.2.2.1 Materials

The material selected for this investigation is polypropylene (PP). Granules of polypropylene and chemical structure of the polypropylene are shown in Figs. 6.17 and 6.18.

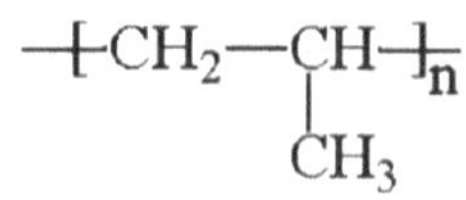

Fig. 6.18 Chemical structure

Table 6.6 Injection molding parameters

S.No.	Parameters	Values
1	Injection time (s)	3
2	Holding time (s)	1.0
3	Injection pressure (MPa)	35
4	Holding pressure (MPa)	40
5	Cooling time (s)	6
6	Injection temperature	
	1. Nozzle zone (°C)	145
	2. Front zone (°C)	150
	3. Center zone (°C)	155
	4. Rear zone (°C)	160

6.2.2.2 Manufacturing Process

The injection molding methods to shape the polymer materials using mold surface or cavity and the ultrasonic welding process for PP joining are discussed in previous injection molding section. The triangular shape of the energy director adopted in this process is shown in Fig. 6.6.

The illustration of the die setup is depicted in Fig. 6.7, and the tabulation of the injection molding processing parameter settings is presented in Table 6.6. The high holding pressure is used to balance the thermal variation at the phase of cooling. After the solidification, the rectangular specimens are subjected to ultrasonic welding trials.

6.2.2.3 Ultrasonic Polymer Welding Process

Ultrasonic welding machine with frequency of 20 kHz and power range of 2000–2400 W is used for this investigation (Fig. 6.8). The characteristics such as patented trigger by power apart from time, process limits, distance controllable provisions, and energy are provided by ultrasonic welding machine. The ultrasonic vibrations welding are used to produce the weld with higher pressure, constant frequency (20 kHz) and low amplitude (20–40 μm). The joining of polymers occurs due to the heat produced due to the stresses, intermolecular friction, and continuous forced contact. In this ultrasonic joining of polymers, heat generation is the critical parameter. The energy director design (EDs) is an important factor during the joining process as material undergoes localized heating at that location.

Table 6.7 Polypropylene parameter levels

Factors	Units	Levels		
		Level 1	Level 2	Level 3
Pressure (P)	Bar	3	4	5
Weld time (T)	s	0.3	0.6	0.9
Amplitude (A)	μm	25	30	35

Fig. 6.19 Ultrasonically welded polypropylene

The pressure, weld time, and amplitude are major parameters influencing the welded specimens. Further, these parameters are designed with design of experiments (DOE) using three different levels. The material characteristics are the key parameter for the welding of amorphous and semicrystalline materials which have diverse melting temperature and properties. Weld area, energy director positions, and schematic of lap joint are shown in Fig. 6.9. The parametric levels for polypropylene joining are presented in Table 6.7. Consequently, the Taguchi method (L27) is adopted to conduct experimental trials. Figure 6.19 shows the welded specimens of polypropylene welded specimens.

6.2.3 *Results and Discussion*

The welded specimens are analyzed with various methods for the mechanical, thermal, and integrity analysis.

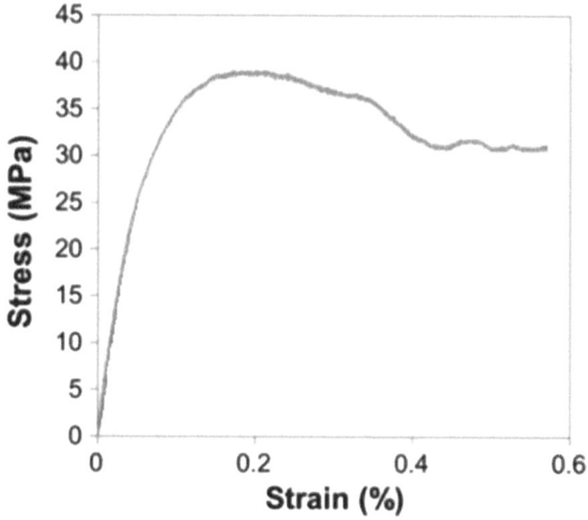

Fig. 6.20 Stress–strain curve for polypropylene material

6.2.3.1 Mechanical Analysis

The weld strength among the materials is determined by the mechanical strength of the welded specimens. The joint is examined by applying the forces in perpendicular direction at the weld interface till the failure occurs as part of tensile testing.

Tensile Test

Stress–strain curve depicts the greater vibration frequency in level 3 amplitude (35 μm) and generates high strength of 38.5 MPa, since frequency generates more vibration among the molecules. Crystalline nature is attained by melting which occurs as a result of greater molecular frictional heat. The rapid melting of crystalline structure will influence more material flow of the material.

The recrystallization of the molten materials occurs during the solidification period, and this recrystallization growth will increase the strength of the weld. The higher weld time of 0.9 s and pressure of 5 bars are set for the level 3. On providing optimal pressure estimated through calculations, the material that undergoes localized melting gradually starts spreading across the surface uniformly. The increased time of weld leads to more melting of material and creation of stronger bond among the specimens as compared to other two levels. The variations are observed for the three levels (level 1, level 2, and level 3) from the stress–strain curve (Fig. 6.20). It may be observed that the curve shows higher plasticity for un-welded specimens while in contrary, the curve representing the welded specimen reveal reduced plasticity and increased strength.

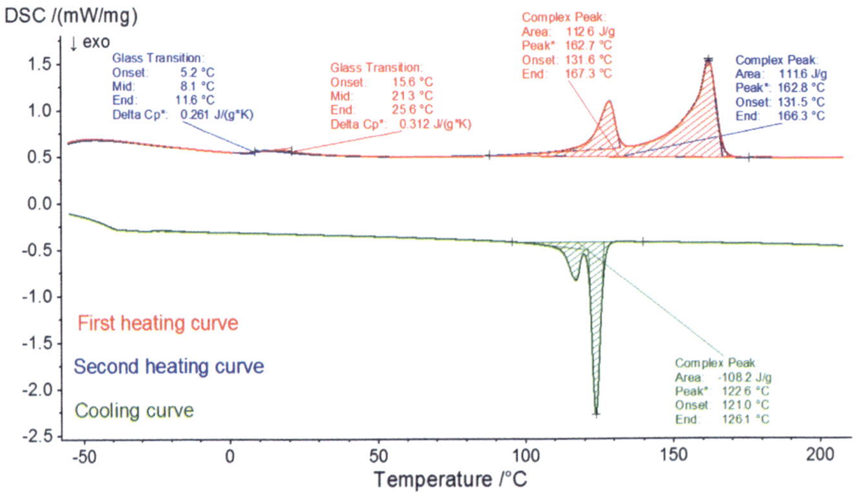

Fig. 6.21 Welded polypropylene material DSC (5 K/min) analysis

6.2.3.2 Thermal Characterization Techniques

Differential Scanning Calorimetric Analysis

Polypropylene Welded Sample Analysis

The observation shows that the endothermic peaks as a result of crystallization melting occur in this phase and no exothermic peaks are noticeable (Fig. 6.21). Cold crystallization is an exothermic process, occurring below the glass transition (Tg) value, whereas molecular mobility is obstructing the crystallization, above the Tg highlighting that crystallization is created at moderately low temperature. Wunderlich (2003) depicted the fact that semicrystalline polymers expose irreversible and reversible melting. The reversible segment during melting responds to remelting, recrystallization, and melting cycles at laminar plane when non-reversible segment features to whole melting of laminar plane.

The melting temperature of molded PP is higher than the melting temperature of welded surface. The initiation of the melting onset and peak values initiated at 131.6 and 167.3 °C, when completing with 162.7 °C Tg values achieved at onset of 15.6 °C, mid value of 21.3 °C, and with the end value of 25.6 °C responding to the heat capacity 0.312 J/g for the first heating time. The observations from the second heating indicate crystalline melting for onset value of 131.6 °C, mid value of 162.7 °C and end values of 11.6 °C respective to Cp values of 0.261 J/g.

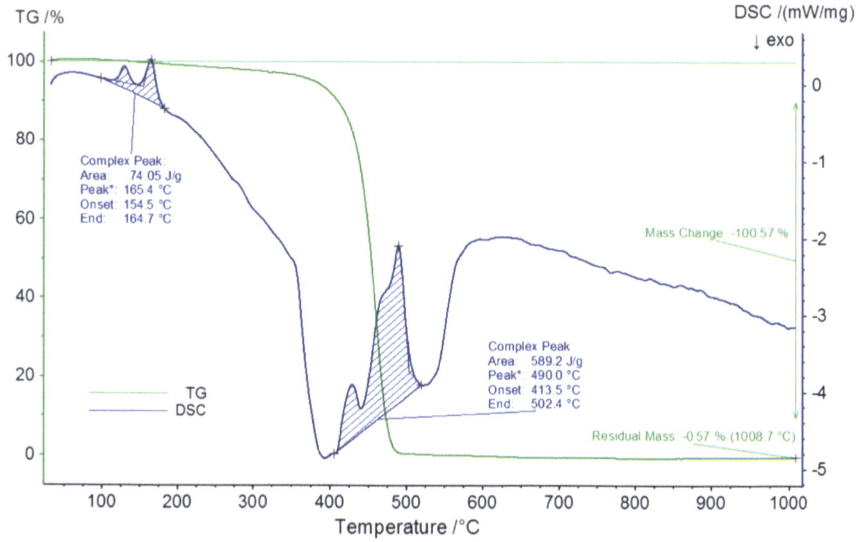

Fig. 6.22 Welded polypropylene material TG analysis

Thermogravimetric Analysis

PP Analysis

The first peak denotes the melting of the polypropylene which begins at 154.5 °C and continues up to 164.7 °C without mass loss (Fig. 6.22). The second peak is decomposition reaction which starts at 165.4–393 °C and the weight is stable in this phase. Consequently, weight loss occurs during the temperature range of 413.5–502.4 °C which is associated with the thermal degradation of the PP main chains. Initiation of the mass loss in the first peak and the complete mass loss in the second peak are observed in the TG curve. In the ultrasonic welding process, the mass loss is not influenced by the vibration and changes in the thermal behavior are observed. From the observation, the original behavior/property of the welded PP specimens is retained during TG analysis where no mass loss is observed, which enhances the physical properties.

6.2.3.3 Integrity Analysis

SEM Analysis for Polypropylene Weldment

SEM analysis reveals that higher amplitude vibration of 35 μm minimizes the voids, as compared to 25 and 30 μm. The higher amplitude vibrations (35 μm) provide few voids in the specimen and increase the bond integrity as observed from the SEM. Nevertheless, voids existed within the limits of the standards (Figs. 6.23, 6.24, and 6.25).

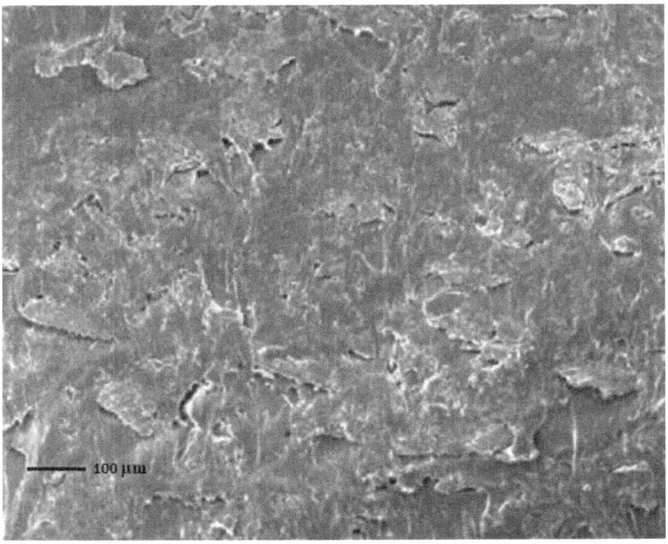

Fig. 6.23 SEM images for polypropylene material (level 1)

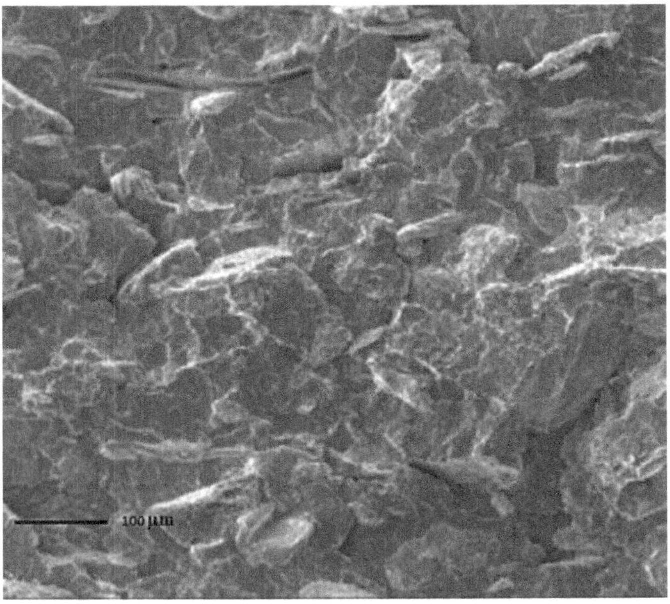

Fig. 6.24 SEM images for polypropylene material (level 2)

Fig. 6.25 SEM images for polypropylene material (level 3)

6.2.4 Conclusions

This investigation focuses on examining the feasibility of ultrasonic lap joining for PP + PP thermoplastics. The weld specimens are fabricated with energy directors embedded on it. The ultrasonic energy used for joining along with sonotrode provides the pressure and vibrations to the specimens. Heat transfer occurs uniformly along the weld interface.

The manufacturer's authorised representative in the EU is Springer Nature Customer Service Centre GmbH, Europaplatz 3, 69115 Heidelberg, Germany. If you have any concerns regarding our products, please contact ProductSafety@springernature.com

Printed and bound by CPI Group (UK) Ltd, Croydon, CR0 4YY
25/09/2026
02222901-0016